CHAPMAN & HALL/CRC APPLIED MATHEMATICS
AND NONLINEAR SCIENCE SERIES

CRC Standard Curves and Surfaces with *Mathematica*®

Second Edition

CHAPMAN & HALL/CRC APPLIED MATHEMATICS AND NONLINEAR SCIENCE SERIES

Series Editors   *Goong Chen and Thomas J. Bridges*

Published Titles

Computing with hp-ADAPTIVE FINITE ELEMENTS: Volume I One and Two Dimensional Elliptic and Maxwell Problems, Leszek Demkowicz

CRC Standard Curves and Surfaces with Mathematica®*: Second Edition,* David H. von Seggern

Exact Solutions and Invariant Subspaces of Nonlinear Partial Differential Equations in Mechanics and Physics, Victor A. Galaktionov and Sergei R. Svirshchevskii

Geometric Sturmian Theory of Nonlinear Parabolic Equations and Applications, Victor A. Galaktionov

Introduction to Fuzzy Systems, Guanrong Chen and Trung Tat Pham

Introduction to Partial Differential Equations with MATLAB®, Matthew P. Coleman

Mathematical Methods in Physics and Engineering with Mathematica, Ferdinand F. Cap

Optimal Estimation of Dynamic Systems, John L. Crassidis and John L. Junkins

Quantum Computing Devices: Principles, Designs, and Analysis, Goong Chen, David A. Church, Berthold-Georg Englert, Carsten Henkel, Bernd Rohwedder, Marlan O. Scully, and M. Suhail Zubairy

Forthcoming Titles

Computing with hp-ADAPTIVE FINITE ELEMENTS: Volume II Frontiers: Three Dimensional Elliptic and Maxwell Problems with Applications, Leszek Demkowicz, Jason Kurtz, David Pardo, Maciej Paszynski, Waldemar Rachowicz, and Adam Zdunek

Introduction to non-Kerr Law Optical Solitions, Anjan Biswas and Swapan Konar

Mathematical Theory of Quantum Computation, Goong Chen and Zijian Diao

Mixed Boundary Value Problems, Dean G. Duffy

Multi-Resolution Methods for Modeling and Control of Dynamical Systems, John L. Junkins and Puneet Singla

Stochastic Partial Differential Equations, Pao-Liu Chow

CHAPMAN & HALL/CRC APPLIED MATHEMATICS
AND NONLINEAR SCIENCE SERIES

CRC Standard Curves and Surfaces with *Mathematica*®
Second Edition

David H. von Seggern

Chapman & Hall/CRC
Taylor & Francis Group

Boca Raton   London   New York

Chapman & Hall/CRC is an imprint of the
Taylor & Francis Group, an informa business

Chapman & Hall/CRC
Taylor & Francis Group
6000 Broken Sound Parkway NW, Suite 300
Boca Raton, FL 33487-2742

© 2007 by Taylor and Francis Group, LLC
Chapman & Hall/CRC is an imprint of Taylor & Francis Group, an Informa business

International Standard Book Number-10: 1-58488-599-8 (Hardcover)
International Standard Book Number-13: 978-1-58488-599-3 (Hardcover)
Library of Congress Card Number 2006011655

Library of Congress Cataloging-in-Publication Data

Von Seggern, David H. (David Henry)
    CRC standard curves and surfaces with Mathematica / David H. von Seggern. -- 2nd ed.
        p. cm. -- (Chapman & Hall/CRC applied mathematics and nonlinear science series)
    Rev. ed. of: CRC standard curves and surfaces. c1993.
    Includes bibliographical references and index.
    ISBN-13: 978-1-58488-599-3 (acid-free paper)
    ISBN-10: 1-58488-599-8 (acid-free paper)
    1. Curves on surfaces--Handbooks, manuals, etc. 2. Mathematica (Computer program lan-guage)--Handbooks, manuals, etc. I. Von Seggern, David H. (David Henry). CRC standard curves and surfaces. II. Title. III. Series.

QA643.V66 2007
516.3'52--dc22   2006011655

Visit the Taylor & Francis Web site at
http://www.taylorandfrancis.com

and the CRC Press Web site at
http://www.crcpress.com

Preface

The second edition of this mathematical reference book (*CRC Standard Curves and Surfaces*) comes 14 years after the first edition in 1992. In fact, there was an earlier volume entitled *CRC Handbook of Mathematical Curves and Surfaces*, published in 1990, so the current volume may be considered a 3rd edition. The motivations for the current edition were several: (1) The *Mathematica*® program has matured considerably since 1992, thus allowing more complex curves and surfaces to be presented. (2) The computing power of desktop computers has increased many-fold, thus allowing formerly prohibitive graphical plots to be computed in a reasonable time; and (3) several important and interesting categories of curves were omitted in the first two editions, largely due to reasons (1) and (2).

There are several major new sections in this volume. One is a new chapter on Green's functions, fundamental to engineering and physics. Plots of Green's functions for basic problems involving the Poisson, wave, diffusion, and Helmholtz equations will be found there. Another new chapter recognizes minimal surfaces as an important and interesting field of mathematics. Knots and links have been added to the chapter on 3-D curves. Archimedean solids, duals of Platonic solids, and stellated forms of these have been added to the chapter on regular polyhedra. Elsewhere, new curves and surfaces have been introduced in almost every chapter. Many chapters have been reorganized and better graphical representations have been produced for many curves and surfaces. The index is considerably expanded to provide the reader with not only a quick way to find curves or surfaces of interest but also to find definitions of common mathematical terms. The format of the book is unchanged from the previous edition, with function definitions on the left-hand pages and corresponding function plots on the right-hand pages, thus maintaining the easy reference-like character of the volume.

The author recognizes that the value of the accompanying *Mathematica* notebooks on CD is undoubtedly much greater than that of the notebooks offered separately with the 1992 edition. A very beneficial advance has been in that *Mathematica* notebook files are now system-independent, working equally well on Macintosh, Windows, and UNIX desktop computers and being interchangeable among them. This enables the notebooks to reach a much wider audience than with the 1992 edition. The notebooks themselves have been upgraded considerably with: (1) more uniform formatting, (2) often more complete documentation on particular curves and surfaces than in the book itself, (3) explanation of the plotting algorithm used if other than by a simple *Mathematica* function, and (4) clearer and more explicit designation of variable parameters such that the users can easily adjust the curve or surface plots to their needs. The notebooks contain the code to construct plots of all the functions in the book; such a unified and coherent body of code can probably not be found elsewhere, while for some individual curves or surfaces it may not exist elsewhere in any readily obtainable form.

In preparing the latest edition, the author has benefited from people, too numerous to mention here, who have communicated by letter or email concerning improvements, corrections, and possible additions; the author wishes to extend his appreciation to these individuals.

The author further wishes to thank the *Mathematica* developers for enabling this third edition with the many new and useful features of the program and for providing stimulus in conferences, in newsletters, and in a rich and extensive website (in particular, http://mathworld.wolfram.com/). The author is indebted to Robert Stern, the mathematics editor of Taylor & Francis, for encouraging and facilitating this latest edition of the work.

Author

David H. Von Seggern, PhD, worked for Teledyne Geotech from 1967 to 1982 in Alexandria, Virginia almost exclusively on analysis of seismic data related to underground nuclear explosions. This effort was supported by the Air Force Office of Scientific Research (AFOSR) and by the Defense Advanced Research Projects Agency (DARPA). His research there addressed detection and discrimination of explosions, physics of the explosive source, explosive yield estimation, wave propagation, and application of statistical methods. Dr. Von Seggern received his PhD at Pennsylvania State University in 1982. He followed that with a 10-year position in geophysics research at Phillips Petroleum Company where he became involved with leading-edge implementation of seismic imaging of oil and gas prospects and with seismic-wave modeling. In 1992, Dr. Von Seggern assumed the role of Seismic Network Manager at the University of Nevada for the Yucca Mountain Project seismic studies. In this capacity, Dr. Von Seggern continued to investigate detection and location of seismic events, elastic wave propagation, and seismic source properties. Dr. Von Seggern retired from full-time work in September 2005 and now pursues various seismological studies as emeritus faculty at the University of Nevada.

Contents

1

Introduction

1.1 Concept of a Curve

Let E^n be the Euclidean space of dimension n. According to this definition, E^1 is a line, E^2 is a plane, and E^3 is a volume. A curve in n-space is defined as the set of points that result when a mapping from E^1 to E^n is performed. In this reference work, only curves in E^2 and E^3 will be considered. Let t represent the independent variable in E^1. An E^2 curve is then given by

$$x = f(t), \ y = g(t)$$

and an E^3 curve by

$$x = f(t), \ y = g(t), \ z = h(t),$$

where f, g, and h mean "function of." The domain of t is usually $(0, 2\pi)$, $(-\infty, \infty)$, or $(0, \infty)$. These are the *parametric representations* of a curve. However, in E^2 curves are commonly expressed as

$$y = f(x)$$

or as

$$f(x, y) = 0,$$

which are the explicit and implicit forms, respectively. The explicit form is readily reducible from the parametric form when $x = f(t) = t$ in E^2 and when $x = f(t) = t$ and $y = g(t) = t$ in E^3. The implicit form of a curve will often comprise more points than a corresponding explicit form. For example, $y^2 - x = 0$ has two ranges in y, one positive and one negative, although the explicit form derived from solving the above equation gives $y = \sqrt{x}$ for which the range of y is positive only.

Generally, the definition of a curve imposes a *smoothness* criterion,[1] meaning that the trace of the curve has no abrupt changes of direction (continuous first derivative). For purposes of this reference work, a broader definition of curve is proposed. Here, a curve may be composed of smooth branches, each satisfying the above definition, provided that the intervals over which the curve branches are distinctly defined and are contiguous. This definition will encompass forms such as polygons or sawtooth functions.

1.2   Concept of a Surface

This reference work defines surfaces as existing only in E^3. Therefore, a surface is defined as the mapping from E^2 to E^3 according to

$$x = f(s,t), \; y = g(s,t), \; z = h(s,t).$$

As for curves, the conversion from this parametric form to more common forms, such as

$$z = f(x,y)$$

or

$$f(x,y,z) = 0,$$

may not be possible in some cases. Again, a smoothness criterion[1] is desirable; however, the generalized definition of a surface only requires that this smoothness criterion be satisfied piecewise for all distinct mappings of the (s, t) plane over which the surface is defined. These generalized surfaces are termed *manifolds*. Cubes are examples of surfaces that may be defined in this deterministic manner.

1.3   Coordinate Systems

The number of available coordinate systems is large for representing curves and even larger for representing surfaces. However, to maintain uniformity of presentation throughout this volume, only the following will be used:

2-D	3-D
Cartesian, polar	Cartesian, cylindrical, spherical

The term *parametric* is often used as though it were a coordinate system, but it is a representation of coordinates in terms of an additional independent parameter that is not itself a coordinate of the E^n space in which the curve or surface exists.

1.3.1   Cartesian Coordinates

The Cartesian coordinate system is illustrated in Figure 1.1 for two dimensions. This is the most natural—but not always the most convenient—system of coordinates for curves in

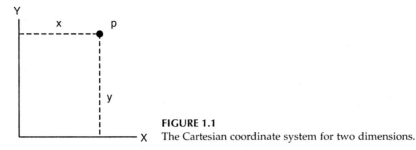

FIGURE 1.1
The Cartesian coordinate system for two dimensions.

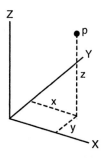

FIGURE 1.2
The Cartesian coordinate system for three dimensions.

two dimensions. Coordinates of a point p are linearly measured along two axes that intersect with a right angle at the origin $(0, 0)$. The Cartesian system is also called the *rectangular coordinates* system. For three dimensions, an additional axis, orthogonal to the other two, is placed as shown in Figure 1.2.

1.3.2 Polar Coordinates

Polar coordinates (r, θ) are defined for two dimensions and are a desirable alternative to Cartesian coordinates when the curve is point symmetric and exists only over a limited domain and range of the variables x and y. As illustrated in Figure 1.3, the coordinate r is the distance of the point p from the origin and the coordinate θ is the counterclockwise angle that the line from the origin to p makes with the horizontal line through the origin to the right. Counterclockwise rotations are measured in positive θ and clockwise rotations are measured in negative θ, relative to this line. Transformations from polar to Cartesian coordinates, and vice-versa, are performed according to:

$$x = r\cos(\theta), \quad y = r\sin(\theta)$$

$$r = (x^2 + y^2)^{1/2}, \quad \theta = \arctan(y/x).$$

1.3.3 Cylindrical Coordinates

Cylindrical coordinates are used in E^3. They combine the (r, θ) polar coordinates of two dimensions with the third coordinate z measured perpendicularly from the $x-y$ plane at (r, θ) to the point p at (r, θ, z) as in Figure 1.4. The normal convention is for z to be positive upward. Transformation from cylindrical to Cartesian coordinates involves only the polar-to-Cartesian transformations given above because the z coordinate is unchanged. Cylindrical coordinates are often appropriate when surfaces are axially symmetric about the z axis; for example, in representing the form $r^2 = z$.

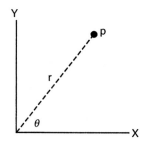

FIGURE 1.3
The polar coordinate system for two dimensions.

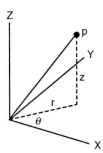

FIGURE 1.4
The cylindrical coordinate system for three dimensions.

1.3.4 Spherical Coordinates

As illustrated in Figure 1.5, let a point in E^3 lie at a radial distance r along a vector from the origin. Project this vector to the $x-y$ plane and let the angle between the vector and its projection be ϕ. Now measure the angle θ of the projected line in the $x-y$ plane as for polar coordinates. Then, (r, θ, ϕ) are the spherical coordinates of p. The transformations from spherical to Cartesian coordinates, and vice-versa, are given by:

$$x = r\cos\theta\sin\phi, \quad y = r\sin\theta\sin\phi, \quad z = r\cos\phi,$$

$$r = (x^2 + y^2 + z^2)^{1/2}, \quad \theta = \arctan(y/x), \quad \phi = \arctan[(x^2 + y^2)^{1/2}/z].$$

Spherical coordinates are often appropriate for surfaces having point symmetry about the origin. The usual coordinates of geography that refer to points on the earth by latitude and longitude are a spherical system.

1.4   Qualitative Properties of Curves and Surfaces

Curves and surfaces exhibit a variety of forms. Particular attributes of form are derivable from the equations themselves and many texts treat these in rigorous detail. The purpose here is not to duplicate such explicit and analytical treatment but rather to present the properties of curves and surfaces in a qualitative manner to which their visible forms are naturally and easily related. Understanding these properties enables one to choose the appropriate curve for a given purpose (for example, data fitting) or to modify, when necessary, an equation given in this volume into one more suitable for a given purpose.

1.4.1   Derivative

A fundamental quantity associated with a curve, or function, is the *derivative*. The *derivative* exists at all continuous points of the curve (except singular points as described in

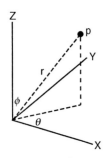

FIGURE 1.5
The spherical coordinate system for three dimensions.

Section 1.4.7). Although the definition of derivative can be made with analytical rigor,[1] in graphical terms the derivative at any point is the slope of the tangent line at that point and is written as $\partial y/\partial x$ for two-dimensional curves. For three-dimensional curves, the tangent line is along the trajectory of the curve, and three such derivatives are possible using the three pairs of (x, y, z) coordinates. Closely associated with the derivative is a curve's *normal* that is the line perpendicular to the tangent. In two dimensions the normal is a single line, but in three dimensions the normal sweeps out a plane perpendicular to the tangent of the curve.

As for curves, the derivative of a surface is a fundamental quantity. The derivative at any continuous point of a surface relates to the tangent plane of the surface at that point. For this plane, three *partial derivatives* exist, written as $\partial y/\partial z$, $\partial z/\partial x$, and $\partial x/\partial y$ (or their inverses), which are the slopes of the lines formed at the intersection of the tangent plane with the $y-z$, $z-x$, and $x-y$ planes, respectively. The normal to the surface at a point is the vector orthogonal to the surface at that point. It is defined at all points for which the surface is smooth by the partial derivatives

$$n_\mathrm{p} = \left[\left(\begin{matrix} \dfrac{\partial y}{\partial s} & \dfrac{\partial y}{\partial t} \\[2mm] \dfrac{\partial z}{\partial s} & \dfrac{\partial z}{\partial t} \end{matrix} \right), \left(\begin{matrix} \dfrac{\partial z}{\partial s} & \dfrac{\partial z}{\partial t} \\[2mm] \dfrac{\partial x}{\partial s} & \dfrac{\partial x}{\partial t} \end{matrix} \right), \left(\begin{matrix} \dfrac{\partial x}{\partial s} & \dfrac{\partial x}{\partial t} \\[2mm] \dfrac{\partial y}{\partial s} & \dfrac{\partial y}{\partial t} \end{matrix} \right) \right]_\mathrm{P}$$

using the parametric representation equations. If the surface can be expressed in the implicit form $f(x, y, z)=0$, then simply

$$n_\mathrm{p} = \left[\dfrac{\partial f}{\partial x}, \ \dfrac{\partial f}{\partial y}, \ \dfrac{\partial f}{\partial z} \right]_\mathrm{P}$$

The above definitions give the (x, y, z) components of the normal vector; it is customary to normalize them to (x', y', z') by dividing them with $(x^2+y^2+z^2)^{1/2}$ so that $x'^2+y'^2+z'^2=1$.

1.4.2 Symmetry

For curves in two dimensions, if

$$y = f(x) = f(-x)$$

holds, then the curve is *symmetric* about the y axis. The curve is *antisymmetric* about the y axis when

$$y = f(x) = -f(-x).$$

A simple example is powers of x: $y=x^n$. If n is even, the curve is symmetric; if n is odd, it is antisymmetric. Antisymmetry is also referred to as "symmetric with respect to the origin" or point symmetry about $(x, y)=(0, 0)$.

For surfaces, three kinds of symmetry exist: point, axial, and plane. A surface has *point symmetry* when

$$z = f(x, y) = -f(-x, -y).$$

Simple examples of point symmetry are spheres or ellipsoids. A surface has *axial symmetry* when

$$z = f(x,y) = f(-x,-y).$$

An example of axial symmetry is a paraboloid. Finally, a surface has *plane symmetry* about the (y, z) plane when

$$z = f(x,y) = f(-x,y).$$

Similarly, symmetry about the (x, z) plane implies

$$z = f(x,y) = f(x,-y).$$

Finally, symmetry about the (x, y) plane is represented by

$$z = f(x,y) = -f(x,y).$$

Examples of plane symmetry include $z=xy^2$ and $z=e^x \cos(y)$.

1.4.3 Extent

The extent of a curve is defined by the *range* (y variation) and *domain* (x variation) of the curve. The extent is *unbounded* if both x and y values can extend to infinity (for example, $y=x^2$). The extent is *semibounded* if either y or x has a bound less than infinity. The transcendental equation $y=\sin(x)$ is such a curve because the range is limited between negative and positive unity. A curve is *fully bounded* if both x and y bounds are less than infinity. A circle is a simple example of this type of extent.

For surfaces the concept of extent can be applied in three dimensions where domain applies to x and y while range applies to z. Surfaces formed by revolution of a curve in the (y, z) or (x, z) plane about the z axis will possess the same extent property that the two-dimensional curve had. For example, an ellipse in the (x, z) plane gives an ellipsoid as the surface of revolution—both have the fully bounded property. Similarly, any surface formed by continuous translation of a two-dimensional curve (for example, a parabolic sheet) will have the same extent property as the original curve.

1.4.4 Asymptotes

The y asymptotes of a curve are defined by

$$y_a = \lim_{x \to \pm\infty} f(x).$$

Although this definition includes asymptotes at infinity, only those with $|y_a| < \infty$ are of interest. Asymptotic values are often crucial in choosing and applying functions. Physically, an equation may or may not properly describe real phenomena, depending on its asymptotic behavior. Note that, even though a curve may be semibounded, its asymptote may not be determinable. An example of a semibounded function with a y asymptote is $y=e^{-x}$ while one without an asymptote is $y=\sin(x)$.

The x asymptotes of a curve may be defined in a similar manner:

$$x_a = \lim_{y \to \pm\infty} f(y)$$

when the function is inverted to give $x = f(y)$. An example of a curve with a finite x asymptote is $y = (c^2 - x^2)^{1/2}$ whose asymptote lies at $x = +c$ or $x = -c$.

In addition, curves may have asymptotes that are any arbitrary lines in the plane, not simply horizontal or vertical lines; and the limiting requirements are similar to the forms given above for horizontal or vertical asymptotes. For instance, the equation $y = x + 1/x$ has $y = x$ as its asymptote.

1.4.5 Periodicity

A curve is defined as *periodic* on x with period X if

$$y = f(x + nX)$$

is constant for all integers n. The transcendental function $y = \sin(ax)$ is an example of a periodic curve. A polar coordinate curve can also be defined as periodic with period α in terms of angle θ if

$$r = f(\theta + n\alpha)$$

is constant for all integers n. An example of such a periodic curve is $r = \cos(4\theta)$ which exhibits eight "petals" evenly spaced around the origin.

Surfaces are periodic on x and y with periods X and Y, respectively, if

$$z = f(x + nX, y + mY)$$

is constant for all integers n and m. A surface also may be periodic in only x or only y. A cylindrical coordinate surface may be periodic with period α in terms of the angle θ if

$$z = f(r, \theta + n\alpha)$$

is constant for all integers n. Another type of periodicity expressible in cylindrical coordinates is in the radial direction with period R, when

$$z = f(r + nR, \theta)$$

is constant for all integers n. An example of such periodicity is given by $z = \cos(2\pi r) \cos(\theta)$, which has a period of $R = 1$.

1.4.6 Continuity

A curve is *continuous* at a point x_0, provided it is defined at x_0, when

$$y^+ = \lim_{x \to x_0^+} f(x)$$

and

$$y^- = \lim_{x \to x_0^-} f(x)$$

are finite and equal. Here, "$+$" and "$-$" refer to approaching x_0 from the right and left, respectively. Discontinuities may be finite or infinite: the former implies $y^+ \neq y^-$ even though they are both finite while the latter implies one or both limits are infinite. For surfaces, the paths to a point $p_0 = (x_0, y_0)$ are infinite in number and continuity exists only if the surface is defined at p_0, and

$$z = \lim_{p \to p_0} f(p)$$

is constant for all possible paths. When the curve or surface is undefined at x_0 or p_0 and the above relations hold, it is said to be discontinuous, but with a *removable discontinuity*. For any points at which the above relations do not hold, the curve or surface is discontinuous, with an *essential discontinuity* at such points. The curve $y = \sin(x)/x$ has a removable discontinuity and is therefore continuous in appearance; $y = 1/x$ has an essential discontinuity at $x = 0$ and is discontinuous in appearance. Curves and surfaces are *differentiable* (meaning the derivative exists) everywhere that they are either continuous or have removable discontinuities.

1.4.7 Singular Points

Curves and surfaces may contain singular points. Writing the function for a two-dimensional curve as

$$f(x, y) = 0.$$

the derivative $\partial y / \partial x$ can be written as

$$\frac{\partial y}{\partial x} = \frac{g(x, y)}{h(x, y)},$$

where g and h are functions of x and y. If, for a given point $p(x, y)$, the functions g and h both vanish, the derivative becomes the indeterminate form $0/0$, and $p(x, y)$ is then a *singular point* of the curve. Singular points imply that two or more branches of the curve meet or cross. If two branches are involved, it is a *double point*; if three are involved, it is a *triple point*, etc. Singularities at triple or higher points are not as commonly encountered as those at double points. Double-point singularities for two-dimensional curves are classified as follows:

1. *Isolated* (or *conjugate*) points are where a single point is disjoint from the remainder of the curve. In this case, the derivative is imaginary.

2. *Node* points are where the two derivatives are real and unequal, such that the curve crosses itself.

3. *Cusp* points are where the derivatives of two arcs become equal and the curve ends at this point. A *cusp of the first kind* involves second derivatives of opposite sign, and a *cusp of the second kind* involves second derivatives of the same sign.

4. *Double cusp* (or *osculation*) points are where the derivatives of two arcs become equal while the two arcs of the curve are continuous along both directions away from such points. Double cusps may also be of the first or second kind, as for single cusps.

Curves having one or more nodes will exhibit *loops* that enclose areas. Curves having osculations may also exhibit loops, on one or both sides of the osculation point.

The concept of singular points is extendable to surfaces. Many surfaces are the result of the revolution of a two-dimensional curve about some line; such surfaces retain the singular points of the curve, except that each such point on the curve, unless on the axis of revolution, becomes a circular ring of singular points centered on the axis of revolution. Singular points also appear on more complicated surfaces, but an analysis of the possibilities is beyond the scope of this volume.

1.4.8 Critical Points

Points of a curve $y=f(x)$ at which the derivative $\partial y/\partial x=0$ are termed *critical points*, of which there are three types:

1. *Maximum* points are where the curve is concave downward and thus the second derivative $\partial^2 y/\partial x^2>0$.
2. *Minimum* points are where the curve is concave upward and thus the second derivative $\partial^2 y/\partial x^2<0$.
3. *Inflection* points are where $\partial^2 y/\partial x^2=0$ and the curve changes its direction of concavity.

For surfaces $z=f(x, y)$, the critical points lie at $\partial z/\partial x=\partial z/\partial y=0$. Maximum and minimum points of surfaces are defined similar to those of curves, except both second derivatives must together be greater than zero or less than zero. In the case that they are of opposite sign, the critical point is termed a *saddle*. Such critical points are *nondegenerate*[2] and are isolated from other critical points. More complicated types of *degenerate* critical points occur for surfaces. Points can be classified as degenerate or nondegenerate, depending on whether the determinant of

$$\begin{pmatrix} \dfrac{\partial^2 z}{\partial x^2} & \dfrac{\partial^2 z}{\partial x \partial y} \\[3ex] \dfrac{\partial^2 z}{\partial x \partial y} & \dfrac{\partial^2 z}{\partial y^2} \end{pmatrix}$$

vanishes or not, respectively. The surface $z=x^2+y^2$ has a single nondegenerate critical point while $z=x^2 y^2$ has two continuous lines of degenerate critical points, intersecting at $(0, 0)$.

1.4.9 Zeroes

The zeroes of a two-dimensional function $f(x)$ occur where $y=f(x)=0$ and are isolated points on the x axis. (For polynomial functions, the zeroes are often referred to as the roots.) Similarly, the zeroes of a three-dimensional function $f(x, y)$ occur where $z=f(x, y)=0$; the loci of these points, however, form one or more distinct, continuous curves in the $x-y$ plane. The zeroes of certain functions are important in characterizing their oscillatory behavior; for example, the function $\sin(x)$. The zeroes of other functions may be unique points of interest in physical applications. Not all functions, as defined, have zeroes; for example, the function $f(x)=2-\cos(x)$ has unity as its lower bound. However, such a function can be translated in one or the other y directions to produce a function having zeroes in addition to all the qualitative properties of the original function.

The definition of the exact zeroes of a function is often difficult and often must be accomplished by numerical methods on a computer. Zeroes of many functions are tabulated in standard references such as Abramowitz.[3]

1.4.10 Integrability

The function $y=f(x)$ defined over the interval $[a, b]$ has the integral

$$I = \int_a^b y\,dx.$$

The integral exists if I converges to a single, bounded value for a given interval and the function is said to be *integrable*. Note that the integral I may exist under two abnormal circumstances:

1. Either a or b, or both, extend to infinity.
2. The function y has an infinite discontinuity at one or both endpoints or at one or more points interior to $[a, b]$.

Under either of these circumstances, the integral is an *improper integral*. Proving the existence of the integral of a given function is not always straightforward; a discussion is beyond the scope of this volume.

Transient functions always have an integral on the interval $[0, \infty]$ and are often given as solutions to physical problems in which the response of a medium to a given input or disturbance is sought. Such responses must possess an integral if the input was finite and measurable. Examples of such functions are $y=e^{-ax}\sin(bx)$ or $y=1/(1+x^2)$.

Surfaces given by $z=f(x, y)$ are integrable when

$$I = \int_a^b \int_c^d z\,dx\,dy$$

exists. Improper integrals of surfaces are defined in the same manner as those of two-dimensional curves. Transient responses exist for three dimensions and are also integrable.

A curve property which has an important consequence for integration is that of even and odd functions. Even functions have $f(x)=f(-x)$, and for such curves

$$I = 2\int_0^a f(x)dx$$

if I exists over $[-a, a]$. For odd functions, $f(x)=f(-x)$, and $I=0$ over any interval $[-a, a]$. This concept can be easily extended to surfaces.

1.4.11 Multiple Values

A curve is *multivalued* if, for a given (x, y), it has two or more distinct values. A simple example is $y^2=x$. Multivalued functions are not integrable in the normal sense, although one or more particular branches of the curve may have well-defined integrals.

Although a curve may be multivalued in its Cartesian-form equation, the polar form of the equation may be single-valued, in the sense that only one value of r exists for each value of angle θ. Compare, for example,

$$(x^2 + y^2)^3 = (x^2 - y^2)^2,$$

which is the equation of a quadrifolium, with its polar equation

$$r = \cos(2\theta).$$

Integrability is affected by the choice of coordinate system; this example shows that when an integral is not defined due to a function being multivalued, it may be well defined when the transformation to polar coordinates is performed and the integral is evaluated along the polar angle θ. Similarly, surfaces may be single-valued or multivalued depending upon whether z takes on one or more values for a given (x, y) point.

1.4.12 Curvature

Given that a unit of length along the curve path is ∂s and that the tangent line changes its direction over ∂s by an angle $\partial\theta$, where θ is the angle of the tangent with the x axis, then the *radius of curvature* is given by

$$\rho = \left| \frac{\partial s}{\partial \theta} \right|.$$

This radius can also be expressed in terms of the derivatives of the curve. If the curve is expressed implicitly as $f(x, y) = 0$, and if f_x and f_y are the first partial derivatives, and f_{xx}, f_{yy}, and f_{xy} are the second partial derivatives, then

$$\rho = \frac{(f_x^2 + f_y^2)^{3/2}}{f_{xx} f_y^2 - 2 f_{xy} f_x f_y + f_{yy} f_x^2}.$$

When the curve is expressed in polar coordinates and the derivatives $\partial r / \partial \theta$ and $\partial^2 r / \partial \theta^2$ are given by r' and r'', respectively, then the radius of curvature is

$$\rho = \frac{(r + r'^2)^{3/2}}{r^2 + 2r'^2 - rr''}.$$

The radius of curvature at lobes of polar curves is of interest to define the "tightness" of the lobes. At the peak of the lobe, $r' = 0$ and $\rho = r^2 / (r - r'')$. This reduces to $\rho = r$ in the case of a circle, for which $r'' = 0$.

Using the same formula as for curves above, curvature of surfaces can be measured along any arbitrary linear arc of the surface made by an intersecting plane, where θ would be the angle of the tangent line relative to the horizontal in the intersecting plane. Thus the curvature of a surface is relative to the perspective it is viewed from.

1.5   Classification of Curves and Surfaces

The family of two-dimensional and three-dimensional curves can be illustrated as in Figure 1.6. This particular schematic reflects the organization of this reference work,

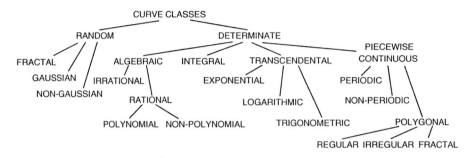

FIGURE 1.6
A classification of curves and surfaces for this handbook.

and every curve that can be traced by a given mathematical equation or given set of mathematical rules can be placed in one of the categories shown. There is a top-level dichotomy between determinate and random curves. A *determinate* curve is one for which the functional relationship between x and y is known everywhere from the equation or set of rules. No realization is required to produce the curve, for it is contained wholly within its defining equations or rules. On the other hand, a *random curve* will have a random factor or term in its mathematical definition such that an actual realization is required to produce the curve, which will differ from any other realization. For example, $y = \sin(x) + w(x)$ where $w(x)$ is a random variable on x, defines a random curve. At the second level in Figure 1.6, the distinction is made between algebraic, transcendental, integral, and nondifferentiable curves as described below.

1.5.1 Algebraic Curves

A *polynomial* is defined as a summation of terms composed of integer powers of x and y. An *algebraic* curve is one whose implicit function

$$f(x, y) = 0$$

is a polynomial in x and y (after rationalization, if necessary). Because a curve is often defined in the explicit form,

$$y = f(x),$$

there is a need to distinguish rational and irrational functions of x. A *rational* function of x is a quotient of two polynomials in x, both having only integer powers. An *irrational* function of x is a quotient of two polynomials, one or both of which contains a term (or terms) with power p/q, where p and q are integers. Irrational functions can be rationalized, but the curves will not be identical before and after rationalization. In general, the rationalized form has more branches. For example, consider $y = \sqrt{x}$, which is rationalized to $y^2 = x$. The former curve has only one branch (for positive y) if a strict definition of the radical is used whereas the latter has two branches, for $y < 0$ and $y > 0$. In this reference work, the rationalized curve will be presented graphically in all cases, even though the equation is printed in its irrational form for simplicity.

Besides simple polynomials, rational functions are often grouped into sets convenient for certain mathematical applications. Examples of such polynomial sets are Chebychev polynomials, Laguerre polynomials, and Bernoulli polynomials. Most polynomial sets have the property of *orthogonality*, meaning that for any two functions f_1 and f_2 of the set,

$$\int w(x)f_1(x)f_2(x)\mathrm{d}x = 0$$

over the defined domain of x for the particular set, where $w(x)$ is a weighting function. This property ensures that the different curves within the set make distinct contributions to the set.

1.5.2 Transcendental Curves

The transcendental curves cannot be expressed as finite polynomials in x and y. These are curves containing one or more of the following forms: exponential (e^x), logarithmic ($\log x$), or trigonometric ($\sin x$, $\cos x$). The hyperbolic functions are often mentioned as part of this group, but they are not distinct because they are forms composed of exponential functions. Any curve expressed as a mixture of transcendental and polynomial is considered to be transcendental. All of the primary transcendental functions can, in fact, be expressed as infinite polynomial series:

$$e^x = \sum_{n=0}^{\infty} \frac{x^n}{n!} \quad (-\infty < x < \infty)$$

$$\cos x = \sum_{n=0}^{\infty} \frac{(-1)^n x^{2n}}{(2n)!} \quad (-\infty < x < \infty)$$

$$\sin x = \sum_{n=0}^{\infty} \frac{(-1)^n x^{2n+1}}{(2n+1)!} \quad (-\infty < x < \infty)$$

$$\log x = 2\sum_{n=1}^{\infty} \frac{1}{2n-1} \left(\frac{x-1}{x+1}\right)^{2n+1} \quad (x > 0)$$

1.5.3 Integral Curves

Certain continuous curves not expressible in algebraic or transcendental forms are familiar mathematical tools. These curves are equal to the integral of algebraic or transcendental curves by definition; examples include Bessel functions, Airy integrals, Fresnel integrals, and the error function. The *integral curve* is given by

$$y(b) = \int_a^b f(x)\,\mathrm{d}x,$$

where the lower limit of integration a is usually a fixed point such as $-\infty$ or 0. Like transcendental curves, these integral curves also have expansions in terms of power series or polynomial series, often making evaluation by computers straightforward.

1.5.4 Piecewise Continuous Functions

Members of the previous classes of curves (algebraic, transcendental, and integral) all have the property that (except at a few points, called singular points) the curve is smooth and differentiable. In the spirit of a broad definition of curve, a class of nondifferentiable curves appears in Figure 1.6. These curves have discontinuity of the first derivative as a basic attribute and are quite often composed of straight-line segments. Such curves include the simple polygonal forms as well as the intricate "regular fractal" curves of Mandelbrot.[4]

1.5.5 Classification of Surfaces

In general, surfaces may follow the same classification scheme as curves (Figure 1.6). Many commonly used surfaces are either rotations of two-dimensional curves about an axis, thus giving axial, or possibly point, symmetry. In this case, the independent variable x of the two-dimensional curve's equation can be replaced with the radial variable $r = (x^2 + y^2)^{1/2}$ to form the equation of the surface. Other commonly used surfaces are merely a continuous translation of a given two-dimensional curve along a straight line. Such surfaces will actually have only one independent variable if a coordinate system having one axis coincident with the straight line is chosen.

If the two independent variables of the explicit equation of the surface, $z = f(x, y)$ are separable in the sense that

$$z = f(x)f(y)$$

then the surface is *orthogonal*. In such a case, the x dependence may fall into one of the classes of Figure 1.6 while the y dependence falls in another. Orthogonal surfaces require fewer operations to evaluate over a grid of the domain of x and y because the defining equation only needs to be evaluated once along the x direction and once along the y direction, with all other points evaluated by simple multiplication of the x and y factors appropriate to each point on the (x, y) plane.

1.6 Basic Curve and Surface Operations

There are many simple operations that can be applied to curves and surfaces to change them. Knowledge of these operations enables one to adapt a given curve or surface to a particular need and to thus extend the curves and surfaces given in this reference volume to a larger set of mathematical forms. Only a few of the most common operations are presented here. Of these, two (translation and rotation) are *homomorphic operations*, in which the curve is preserved and only its position or orientation in space is changed.

1.6.1 Translation

If one of the coordinates x, y, or z of a point is changed according to

$$x' = x + a,$$

$$y' = y + b, \text{ or}$$

$$z' = z + c,$$

then the curve or surface undergoes a *translation* of amount (a, b, c) along the (x, y, z) axes, respectively.

1.6.2 Rotation

In polar coordinates, if the angle θ is changed by a positive amount α,

$$\theta' = \theta + \alpha,$$

the curve undergoes a counterclockwise *rotation* of α degrees. This is convenient for polar coordinates, but the rotation can also be expressed in Cartesian coordinates as

$$x' = x \cos(\alpha) + y \sin(\alpha)$$
$$y' = -x \sin(\alpha) + y \cos(\alpha).$$

In three dimensions, a surface can be rotated about any of the three axes by using these equations on the coordinate pairs (x, y), (y, z), or (x, z), depending on whether the rotation is about the z, x, or y axis, respectively.

1.6.3 Linear Scaling

The relations for *linear scaling* are

$$x' = ax, \; y' = by, \; z' = cz.$$

These stretch the curve or surface by the factors a, b, and c along the respective axes. When polar, cylindrical, or spherical coordinates are used, a similar relation

$$r' = dr$$

stretches or compresses the curve or surface along the radial coordinate by the factor d.

1.6.4 Reflection

A two-dimensional curve has a *reflection* about the x axis caused by letting

$$y' = -y,$$

or about the y axis by letting

$$x' = -x,$$

or through the origin by applying both these equations. In three dimensions, a curve or surface is reflected across the (y, z), (x, z), or (x, y) planes when

$$x' = -x$$
$$y' = -y$$
$$z' = -z,$$

respectively. It can be reflected through the origin when one sets

$$r' = -r$$

in spherical coordinates and mirrored through the z axis when the same operation is made on r for cylindrical coordinates. The application to two-dimensional polar coordinates follows from the cylindrical case.

1.6.5 Rotational Scaling

For two dimensions, let

$$\theta' = c\theta$$

for the polar angle; the polar curve is then stretched or compressed along the angular direction by a factor c in a *rotational scaling*. The same operation can be applied to θ for cylindrical coordinates in three dimensions or to both θ and ϕ for spherical coordinates in three dimensions.

1.6.6 Radial Translation

In two dimensions with polar coordinates, if the radial coordinate is translated according to

$$r' = r + a,$$

then the entire curve moves outward by the amount a from the origin. Note that this operation is not homomorphic like Cartesian translation because the curve is stretched in the angular direction while undergoing the radial translation. This operation can be performed on the radial coordinate of either cylindrical or spherical coordinate systems in three dimensions.

1.6.7 Weighting

In a two-dimensional Cartesian system, let

$$y' = |x|^a y.$$

This operation performs a *weighting* on the curve by the factor $|x|^a$, a symmetric operator. If $a > 0$, the curve is stretched in the y direction by a factor that increases with x; but if $a < 0$, the curve is compressed by a factor that decreases with x. Similar treatments may be performed on surfaces in three dimensions.

1.6.8 Nonlinear Scaling

If in two dimensions, the *nonlinear scaling*,

$$y' = y^a,$$

is performed, the curve is progressively scaled upward or downward in absolute value, according to whether $a > 1$ or $a < 1$, respectively. Note that, if $y < 0$ and $a = 2, 4, 6, \ldots$, then the scaled curve will flip to the opposite side of the x axis. Similar scalings can be made in three dimensions using any of the appropriate coordinate systems.

1.6.9 Shear

A curve undergoes *simple shear* when either all of its x coordinates or all of its y coordinates remain constant while the other set is increased in proportion to x or y, respectively. The general transformations for simple shearing of a two-dimensional curve are

$$x' = x + ay,$$

$$y' = bx + y,$$

The transformations for simple x shear are

$$x' = x + ay$$

$$y' = y$$

and for simple y shear are

$$x' = x$$

$$y' = y + bx$$

Surfaces may be simply sheared along one or two axes with similar transformations. Another special case of shear is termed *pure shear*, and the transformations for a two-dimensional curve are given by

$$x' = kx$$

$$y' = k^{-1}y$$

For surfaces, pure shear will only apply to two of the three coordinate directions, with the remaining one having no change. Pure shear is a special case of linear scaling under this circumstance.

1.6.10 Matrix Method for Transformation

The foregoing transformations can all be expressed in matrix form, which is often convenient for computer algorithms. This is especially true when several transformations are concatenated together, for the matrices can then be simply multiplied together to obtain a single transformation matrix. Given a pair of coordinates (x, y), a matrix transformation to obtain the new coordinates (x', y') is written as

$$(x' \quad y') = (x \quad y) \begin{pmatrix} a & b \\ c & d \end{pmatrix}$$

or explicitly as

$$x' = ax + cy$$

$$y' = bx + dy$$

According to this definition, Table 1.1 lists several of the two-dimensional $x-y$ transformations discussed previously with their corresponding matrix.

Translations cannot be treated with the above matrix definition. An extension is required to produce what is commonly referred to as the *homogeneous coordinate representation* in computer graphics programming. In its simplest form, an additional coordinate of unity is appended to the (x, y) pair to give $(x, y, 1)$. A translation by u and v in the x and y directions is then written using a 3×3 matrix:

$$(x' \quad y' \quad 1) = (x \quad y \quad 1) \begin{pmatrix} 1 & 0 & 0 \\ 0 & 1 & 0 \\ u & v & 1 \end{pmatrix}$$

TABLE 1.1

2-D Transformations

Operation	Matrix	Notes
Rotation	$\begin{pmatrix} \cos\alpha & \sin\alpha \\ -\sin\alpha & \cos\alpha \end{pmatrix}$	α is the counterclockwise angle in the $x-y$ plane
Linear scaling	$\begin{pmatrix} a & 0 \\ 0 & b \end{pmatrix}$	
Reflection	$\begin{pmatrix} \pm 1 & 0 \\ 0 & \pm 1 \end{pmatrix}$	Use $+$ or $-$ according to desired reflection
Weighting	$\begin{pmatrix} 1 & 0 \\ 0 & x^a \end{pmatrix}$	
Nonlinear scaling	$\begin{pmatrix} 1 & 0 \\ 0 & y^a \end{pmatrix}$	
Simple shear	$\begin{pmatrix} 1 & a \\ b & 1 \end{pmatrix}$	Either a or b is zero; former gives simple x shear and latter gives simple y shear
Rotational scaling	$\begin{pmatrix} 1 & 0 \\ 0 & a \end{pmatrix}$	Use with (r,θ) coordinates

where explicitly,

$$x' = x + u$$

$$y' = y + v$$

$$1 = 1$$

With this representation, a radial translation by s units of a curve given in (r, θ) coordinates is effected by

$$(r' \quad \theta \quad 1) = (r \quad \theta \quad 1) \begin{pmatrix} 1 & 0 & 0 \\ 0 & 1 & 0 \\ s & 0 & 1 \end{pmatrix}$$

such that $r' = r + s$ and θ is unchanged.

In three dimensions similar transformations exist (Table 1.2), mostly being simple extensions of those given in Table 1.1.

1.7 Method of Presentation

This reference work is basically intended to be illustrative; therefore, all functions, whether curves or surfaces, presented in this volume will have an accompanying plot showing the form of the function. The plot will, in all cases, be on the right-hand page

TABLE 1.2

3-D Transformations

Operation	Matrix	Notes
Rotation	$\begin{pmatrix} \cos\beta\cos\gamma & \cos\gamma\sin\alpha\sin\beta + \cos\alpha\sin\gamma & -\cos\alpha\cos\gamma\sin\beta + \sin\alpha\sin\gamma \\ -\cos\beta\sin\gamma & \cos\alpha\cos\gamma - \sin\alpha\sin\beta\sin\gamma & \cos\gamma\sin\alpha + \cos\alpha\sin\beta\sin\gamma \\ \sin\beta & -\cos\beta\sin\alpha & \cos\alpha\cos\beta \end{pmatrix}$	α, β, γ are the counterclockwise rotations about each axis, looking from the positive side
Linear scaling	$\begin{pmatrix} a & 0 & 0 \\ 0 & b & 0 \\ 0 & 0 & c \end{pmatrix}$	
Reflection	$\begin{pmatrix} \pm1 & 0 & 0 \\ 0 & \pm1 & 0 \\ 0 & 0 & \pm1 \end{pmatrix}$	Use + or − according to desired reflection
Weighting	$\begin{pmatrix} 1 & 0 & 0 \\ 0 & 1 & 0 \\ 0 & 0 & x^a y^b \end{pmatrix}$	
Nonlinear scaling	$\begin{pmatrix} 1 & 0 & 0 \\ 0 & 1 & 0 \\ 0 & 0 & z^a \end{pmatrix}$	
Simple shear	$\begin{pmatrix} 1 & 0 & 0 \\ 0 & 1 & 0 \\ a & 0 & 1 \end{pmatrix}$ or $\begin{pmatrix} 1 & 0 & 0 \\ a & 1 & 0 \\ 0 & 0 & 1 \end{pmatrix}$	Gives simple x shear, depending whether done along y or z direction. Similar expressions hold for simple y or z shear
Rotational scaling	$\begin{pmatrix} 1 & 0 & 0 \\ 0 & a & 0 \\ 0 & 0 & b \end{pmatrix}$	Use with (r, θ, ϕ) coordinates

while the equation will be on the facing left-hand page. Curves and surfaces and their plots are numbered for easy reference and grouped according to type. Wherever popular names exist for certain curves or surfaces, they are placed with the equations themselves. Only basic explanatory information is provided with each curve, if needed. The interested reader can consult textbooks, or the World Wide Web resources for further information. The accompanying *Mathematica*® notebook often provides details on the construction and behavior of particular curves and surfaces.

1.7.1 Equations

The equation of each algebraic or transcendental curve will be given in the explicit form $y=f(x)$ or $r=f(\theta)$ wherever possible; similarly, surfaces will be given as $z=f(x, y)$ or $r=f(\theta, z)$ or $r=f(\theta, \phi)$. Whenever polar, cylindrical, or spherical coordinate forms are used, the equation is also written in Cartesian coordinates, if possible. Because some curves and surfaces are not amenable to explicit forms, the parametric equations will be used as the alternative when necessary. In either case, whether explicit or parametric, the implicit functional form will also be given, if derivable. The explicit or parametric form is needed to evaluate the curve or surface on a computer while the implicit form enables one to determine the degree of the equation (if algebraic) and also easily determine the derivatives in some cases. Notes pertinent to evaluation are given whenever they may help to understand the figures better.

For integral curves and surfaces, the equation will be given as the integral $y=\int f(x)$ or $z=\int f(x, y)$. Most of the integral forms have commonly used names (for example, "Bessel functions"). Other curves or surfaces in this reference work are expressed not by single equations, but rather by some set of mathematical rules. The method of presentation will vary in these cases, always with the objective of providing the reader with a means of easily constructing the curve or surface by machine computation.

1.7.2 Plots

Plots of two-dimensional curves will be done on the (x, y) plane, with the x and y axes being horizontal and vertical, respectively. The domain of x and the range of y, unless otherwise stated, will be -1 to $+1$; the variable form of the curve will be adequately illustrated by a suitable choice of x and y scaling factors and of the constants in the equation. For example, the curve $y=\sin(x)$ can be illustrated for a domain larger than ±1 by actually plotting $y=\sin(ax)$, with $a>1$, while still letting x vary between -1 and $+1$. Similarly, the range of y can be limited to ±1 by plotting $y=cf(x)$ where the constant c is suitably chosen. Three-dimensional curves and surfaces will have the additional z axis, also from -1 to $+1$, and will be plotted in a projection that satisfactorily illustrates the form of each function. Simple equations will be illustrated by a single plotted curve or surface while more complicated equations may have two or more such plots with different constants to indicate the variation possible in a family of curves or surfaces.

In the case of curves that are unbounded in y (for example, $y=1/x$), the evaluation algorithm computes and plots the curve up to exactly $y=+1$ or $y=-1$. Curves expressed in polar coordinates (r, θ) are similarly truncated at $r=1$ in the case that r is unbounded. The implicit form of a curve will often comprise more points than a corresponding explicit form. For example, $y^2-x=0$ has two ranges in y, one positive and one negative, although the explicit form derived from solving the above equation gives $y=\sqrt{x}$ for which the range of y is positive only; in such cases, both the positive and negative range of y are plotted.

References

1. R.C. Buck. 1965. *Advanced Calculus*, New York: McGraw-Hill, Ch. 5, pp. 238–241.
2. T. Poston and I. Stewart. 1978. *Catastrophe Theory and Its Applications*, New York: Pitman.
3. M. Abramowitz, ed. 1974. *Handbook of Mathematical Functions, with Formulas, Graphs, and Mathematical Tables*, New York: Dover.
4. B.B. Mandelbrot. 1983. *The Fractal Geometry of Nature*, San Francisco: W.H. Freeman.

2

Algebraic Curves

The curves in this chapter are found in elementary algebra texts or tables of integrals and are likely familiar. Many have acquired common names in the mathematical literature and these names are included wherever appropriate. The final two sections deal with curves more easily expressed in polar coordinates or parametrically; this facilitates computation of the curves more than the form $y=f(x)$, especially when curves are multiple-valued. For curves involving radicals, both the positive and negative branches are plotted to show the symmetry.

2.1 Functions with $x^{n/m}$

2.1.1 $y = cx^n$ $y - cx^n = 0$

      1. $c=1.0$, $n=1$ (*linear*)
      2. $c=1.0$, $n=2$ (*quadratic or simple parabola*)
      3. $c=1.0$, $n=3$ (*cubic*)
      4. $c=1.0$, $n=4$ (*quartic*)
      5. $c=1.0$, $n=5$ (*quintic*)
      6. $c=1.0$, $n=6$ (*sextic*)

2.1.2 $y = c/x^n$ $yx^n - c = 0$

      1. $c=0.01$, $n=1$ (*hyperbola*)
      2. $c=0.01$, $n=2$
      3. $c=0.01$, $n=3$
      4. $c=0.01$, $n=4$
      5. $c=0.01$, $n=5$
      6. $c=0.01$, $n=6$

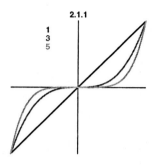

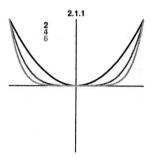

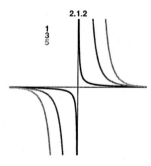

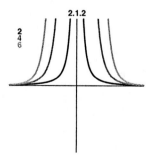

2.1.3 $y = cx^{n/m}$ $y - cx^{n/m} = 0$

>   1. $c=1.0, n=1, m=4$
>   2. $c=1.0, n=1, m=2$
>   3. $c=1.0, n=3, m=4$
>   4. $c=1.0, n=5, m=4$
>   5. $c=1.0, n=3, m=2$ *(semicubical parabola)*
>   6. $c=1.0, n=7, m=4$

>   7. $c=1.0, n=1, m=3$
>   8. $c=1.0, n=2, m=3$ *(cusp catastrophe)*
>   9. $c=1.0, n=4, m=3$
>   10. $c=1.0, n=5, m=3$

2.1.4 $y = c/x^{n/m}$ $yx^{n/m} - c = 0$

>   1. $c=0.1, n=1, m=4$
>   2. $c=0.1, n=1, m=2$
>   3. $c=0.1, n=3, m=4$
>   4. $c=0.1, n=5, m=4$
>   5. $c=0.1, n=3, m=2$
>   6. $c=0.1, n=7, m=4$

>   7. $c=0.1, n=1, m=3$
>   8. $c=0.1, n=2, m=3$
>   9. $c=0.1, n=4, m=3$
>   10. $c=0.1, n=5, m=3$

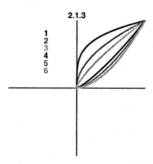

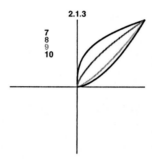

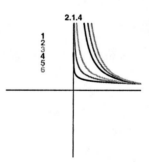

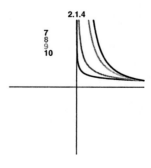

2.2 Functions with x^n and $(a+bx)^m$

2.2.1 $y = c(a + bx)$ $y - bcx - ac = 0$

  1. $a=0.5, b=0.5, c=1.0$
  2. $a=0.5, b=1.0, c=1.0$
  3. $a=0.5, b=2.0, c=1.0$

2.2.2 $y = c(a + bx)^2$ $y - cb^2x^2 - 2abcx - a^2c = 0$

  1. $a=0.5, b=0.5, c=1.0$
  2. $a=0.5, b=1.0, c=1.0$
  3. $a=0.5, b=2.0, c=1.0$

2.2.3 $y = c(a + bx)^3$ $y - b^3cx^3 - 3ab^2cx^2 - 3a^2bcx - a^3c = 0$

  1. $a=0.5, b=0.5, c=1.0$
  2. $a=0.5, b=1.0, c=1.0$
  3. $a=0.5, b=2.0, c=1.0$

2.2.4 $y = cx(a + bx)$ $y - bcx^2 - acx = 0$

  1. $a=0.5, b=0.5, c=1.0$
  2. $a=0.5, b=1.0, c=1.0$
  3. $a=0.5, b=2.0, c=1.0$

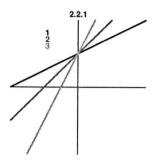

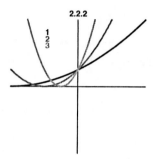

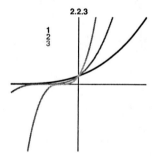

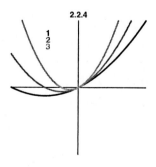

2.2.5 $\qquad y = cx(a + bx)^2 \qquad y - b^2cx^3 - 2abcx^2 - a^2cx = 0$

       1. $a=0.5, b=0.5, c=1.0$
       2. $a=0.5, b=1.0, c=1.0$
       3. $a=0.5, b=2.0, c=1.0$

2.2.6 $\qquad y = cx(a + bx)^3 \qquad y - b^3cx^4 - 3ab^2cx^3 - 3a^2bcx^2 - a^3cx = 0$

       1. $a=0.5, b=0.5, c=1.0$
       2. $a=0.5, b=1.0, c=1.0$
       3. $a=0.5, b=2.0, c=1.0$

2.2.7 $\qquad y = cx^2(a + bx) \qquad y - bcx^3 - acx^2 = 0$

       1. $a=0.5, b=0.5, c=1.0$
       2. $a=0.5, b=1.0, c=1.0$
       3. $a=0.5, b=2.0, c=1.0$

2.2.8 $\qquad y = cx^2(a + bx)^2 \qquad y - b^2cx^4 - 2abcx^3 - a^2cx^2 = 0$

       1. $a=0.5, b=0.5, c=1.0$
       2. $a=0.5, b=1.0, c=1.0$
       3. $a=0.5, b=2.0, c=1.0$

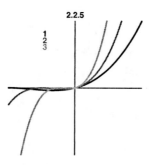

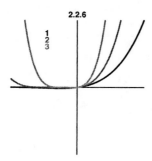

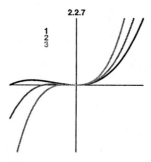

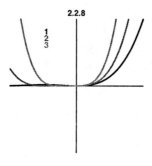

2.2.9 $y = cx^2(a + bx)^3$ $y - b^3cx^5 - 3ab^2cx^4 - 3a^2bcx^3 - a^3cx^2 = 0$

   1. $a=0.5, b=0.5, c=1.0$
   2. $a=0.5, b=1.0, c=1.0$
   3. $a=0.5, b=2.0, c=1.0$

2.2.10 $y = cx^3(a + bx)$ $y - bcx^4 - acx^3 = 0$

   1. $a=0.5, b=0.5, c=1.0$
   2. $a=0.5, b=1.0, c=1.0$
   3. $a=0.5, b=2.0, c=1.0$

2.2.11 $y = cx^3(a + bx)^2$ $y - b^2cx^5 - 2abcx^4 - a^2cx^3 = 0$

   1. $a=0.5, b=0.5, c=1.0$
   2. $a=0.5, b=1.0, c=1.0$
   3. $a=0.5, b=2.0, c=1.0$

2.2.12 $y = cx^3(a + bx)^3$ $y - b^3cx^6 - 3ab^2cx^5 - 3a^2bcx^4 - a^3cx^3 = 0$

   1. $a=0.5, b=0.5, c=1.0$
   2. $a=0.5, b=1.0, c=1.0$
   3. $a=0.5, b=2.0, c=1.0$

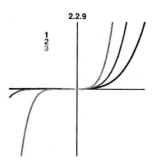

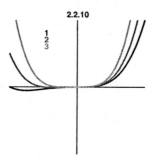

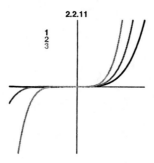

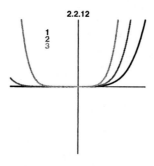

2.2.13 $y = c/(a + bx)$ $ay + bxy - c = 0$

      1. $a=1.0$, $b=2.0$, $c=0.02$
      2. $a=1.0$, $b=3.0$, $c=0.02$
      3. $a=1.0$, $b=4.0$, $c=0.02$

2.2.14 $y = c/(a + bx)^2$ $a^2y + 2abxy + b^2x^2y - c = 0$

      1. $a=1.0$, $b=2.0$, $c=0.02$
      2. $a=1.0$, $b=3.0$, $c=0.02$
      3. $a=1.0$, $b=4.0$, $c=0.02$

2.2.15 $y = c/(a + bx)^3$ $a^3y + 2a^2bxy + 2ab^2x^2y + b^3x^3y - c = 0$

      1. $a=1.0$, $b=2.0$, $c=0.02$
      2. $a=1.0$, $b=3.0$, $c=0.02$
      3. $a=1.0$, $b=4.0$, $c=0.02$

2.2.16 $y = cx/(a + bx)$ $ay + bxy - cx = 0$

      1. $a=1.0$, $b=2.0$, $c=0.1$
      2. $a=1.0$, $b=3.0$, $c=0.1$
      3. $a=1.0$, $b=4.0$, $c=0.1$

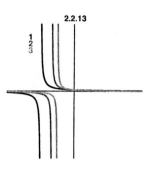

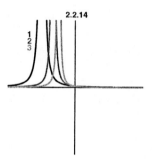

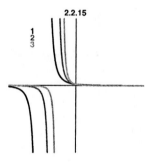

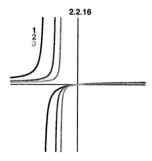

2.2.17 $y = cx/(a + bx)^2$ $a^2y + 2abxy + b^2x^2y - cx = 0$

      1. $a=1.0$, $b=2.0$, $c=0.02$
      2. $a=1.0$, $b=3.0$, $c=0.02$
      3. $a=1.0$, $b=4.0$, $c=0.02$

2.2.18 $y = cx/(a + bx)^3$ $a^3y + 3a^2bxy + 3ab^2x^2y + b^3x^3y - cx = 0$

      1. $a=1.0$, $b=2.0$, $c=0.01$
      2. $a=1.0$, $b=3.0$, $c=0.01$
      3. $a=1.0$, $b=4.0$, $c=0.01$

2.2.19 $y = cx^2/(a + bx)$ $ay + bxy - cx^2 = 0$

      1. $a=1.0$, $b=2.0$, $c=0.2$
      2. $a=1.0$, $b=3.0$, $c=0.2$
      3. $a=1.0$, $b=4.0$, $c=0.2$

2.2.20 $y = cx^2/(a + bx)^2$ $a^2y + 2abxy + b^2x^2y - cx^2 = 0$

      1. $a=1.0$, $b=2.0$, $c=0.2$
      2. $a=1.0$, $b=3.0$, $c=0.2$
      3. $a=1.0$, $b=4.0$, $c=0.2$

2.2.17

1
2
3

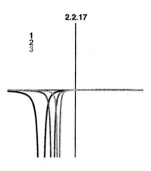

2.2.18

1
2
3

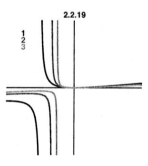

2.2.19

1
2
3

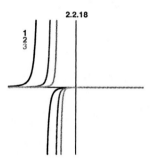

2.2.20

1
2
3

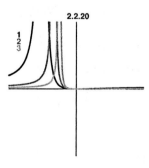

2.2.21 $y = cx^2/(a + bx)^3$ $a^3y + 3a^2bxy + 3ab^2x^2y + b^3x^3y - cx^2 = 0$

    1. $a=1.0$, $b=2.0$, $c=0.02$
    2. $a=1.0$, $b=3.0$, $c=0.02$
    3. $a=1.0$, $b=4.0$, $c=0.02$

2.2.22 $y = cx^3/(a + bx)$ $ay + bxy - cx^3 = 0$

    1. $a=1.0$, $b=2.0$, $c=1.0$
    2. $a=1.0$, $b=3.0$, $c=1.0$
    3. $a=1.0$, $b=4.0$, $c=1.0$

2.2.23 $y = cx^3/(a + bx)^2$ $a^2y + 2abxy + b^2x^2y - cx^3 = 0$

    1. $a=1.0$, $b=2.0$, $c=0.2$
    2. $a=1.0$, $b=3.0$, $c=0.2$
    3. $a=1.0$, $b=4.0$, $c=0.2$

2.2.24 $y = cx^3/(a + bx)^3$ $a^3y + 3a^2bxy + 3ab^2x^2y + b^3x^3y - cx^3 = 0$

    1. $a=1.0$, $b=2.0$, $c=0.1$
    2. $a=1.0$, $b=3.0$, $c=0.1$
    3. $a=1.0$, $b=4.0$, $c=0.1$

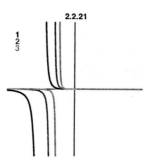

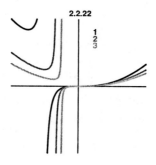

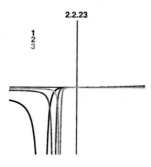

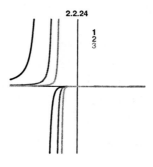

2.2.25 $y = c(a + bx)/x$ $xy - bcx - ca = 0$

      1. $a=1.0$, $b=2.0$, $c=0.04$
      2. $a=1.0$, $b=4.0$, $c=0.04$
      3. $a=1.0$, $b=6.0$, $c=0.04$

2.2.26 $y = c(a + bx)^2/x$ $xy - b^2cx^2 - 2abcx - a^2c = 0$

      1. $a=1.0$, $b=2.0$, $c=0.04$
      2. $a=1.0$, $b=4.0$, $c=0.04$
      3. $a=1.0$, $b=6.0$, $c=0.04$

2.2.27 $y = c(a + bx)^3/x$ $xy - b^3cx^3 - 3ab^2cx^2 - 3a^2bcx - a^3c = 0$

      1. $a=1.0$, $b=2.0$, $c=0.02$
      2. $a=1.0$, $b=4.0$, $c=0.02$
      3. $a=1.0$, $b=6.0$, $c=0.02$

2.2.28 $y = c(a + bx)/x^2$ $x^2y - bcx - ca = 0$

      1. $a=1.0$, $b=2.0$, $c=0.04$
      2. $a=1.0$, $b=4.0$, $c=0.04$
      3. $a=1.0$, $b=6.0$, $c=0.04$

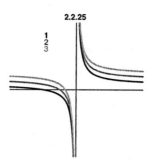

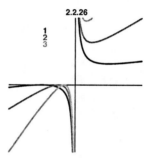

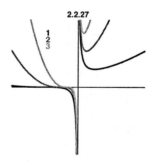

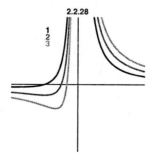

2.2.29 $y = c(a + bx)^2/x^2$ $x^2y - b^2cx^2 - 2abcx - a^2c = 0$

> 1. $a=1.0$, $b=2.0$, $c=0.01$
> 2. $a=1.0$, $b=4.0$, $c=0.01$
> 3. $a=1.0$, $b=6.0$, $c=0.01$

2.2.30 $y = c(a + bx)^3/x^2$ $x^2y - b^3cx^3 - 3ab^2cx^2 - 3a^2bcx - a^3c = 0$

> 1. $a=1.0$, $b=2.0$, $c=0.003$
> 2. $a=1.0$, $b=4.0$, $c=0.003$
> 3. $a=1.0$, $b=6.0$, $c=0.003$

2.2.31 $y = c(a + bx)/x^3$ $x^3y - bcx - ca = 0$

> 1. $a=1.0$, $b=2.0$, $c=0.02$
> 2. $a=1.0$, $b=4.0$, $c=0.02$
> 3. $a=1.0$, $b=6.0$, $c=0.02$

2.2.32 $y = c(a + bx)^2/x^3$ $x^3y - b^2cx^2 - 2abcx - a^2c = 0$

> 1. $a=1.0$, $b=2.0$, $c=0.01$
> 2. $a=1.0$, $b=4.0$, $c=0.01$
> 3. $a=1.0$, $b=6.0$, $c=0.01$

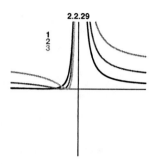

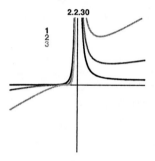

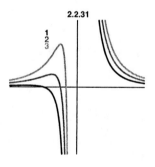

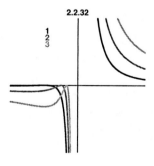

2.2.33 $y = c(a + bx)^3/x^3$ $x^3y - b^3cx^3 - 3ab^2cx^2 - 3a^2bcx - a^3c = 0$

       1. $a=1.0$, $b=2.0$, $c=0.002$
       2. $a=1.0$, $b=4.0$, $c=0.002$
       3. $a=1.0$, $b=6.0$, $c=0.002$

2.3 Functions with $a^2 + x^2$ and x^m

2.3.1 $y = c/(a^2 + x^2)$ $a^2y + x^2y - c = 0$

Special case: $c=a^3$ gives *Witch of Agnesi*

       1. $a=0.2$, $c=0.04$
       2. $a=0.5$, $c=0.04$
       3. $a=0.8$, $c=0.04$

2.3.2 $y = cx/(a^2 + x^2)$ $a^2y + x^2y - cx = 0$

Serpentine

       1. $a=0.2$, $c=0.3$
       2. $a=0.5$, $c=0.3$
       3. $a=0.8$, $c=0.3$

2.3.3 $y = cx^2/(a^2 + x^2)$ $a^2y + x^2y - cx^2 = 0$

       1. $a=0.2$, $c=1.0$
       2. $a=0.5$, $c=1.0$
       3. $a=0.8$, $c=1.0$

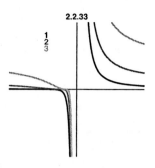

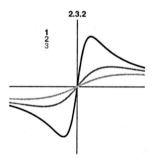

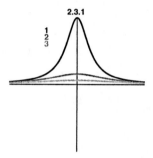

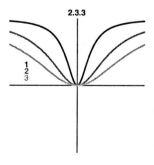

2.3.4 $y = cx^3/(a^2 + x^2)$ $a^2y + x^2y - cx^3 = 0$

        1. $a=0.2$, $c=1.0$
        2. $a=0.5$, $c=1.0$
        3. $a=0.8$, $c=1.0$

2.3.5 $y = c/[x(a^2 + x^2)]$ $a^2 + x^3y - c = 0$

        1. $a=0.2$, $c=0.02$
        2. $a=0.5$, $c=0.02$
        3. $a=0.8$, $c=0.02$

2.3.6 $y = c/[x^2(a^2 + x^2)]$ $a^2x^2y + x^4y - c = 0$

        1. $a=0.2$, $c=0.02$
        2. $a=0.5$, $c=0.02$
        3. $a=0.8$, $c=0.02$

2.3.7 $y = cx(a^2 + x^2)$ $y - a^2cx - cx^3 = 0$

        1. $a=0.2$, $c=1.0$
        2. $a=0.5$, $c=1.0$
        3. $a=0.8$, $c=1.0$

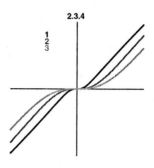

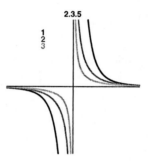

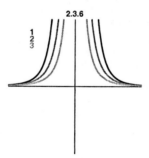

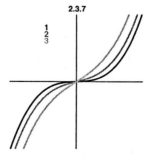

2.3.8 $y = cx^2(a^2 + x^2)$ $y - a^2cx^2 - cx^4 = 0$

    1. $a=0.2, c=1.0$
    2. $a=0.5, c=1.0$
    3. $a=0.8, c=1.0$

2.4 Functions with $a^2 - x^2$ and x^m

2.4.1 $y = c/(a^2 - x^2)$ $a^2y - x^2y - c = 0$

    1. $a=0.2, c=0.03$
    2. $a=0.5, c=0.03$
    3. $a=0.8, c=0.03$

2.4.2 $y = cx/(a^2 - x^2)$ $a^2y - x^2y - cx = 0$

    1. $a=0.2, c=0.1$
    2. $a=0.5, c=0.1$
    3. $a=0.8, c=0.1$

2.4.3 $y = cx^2/(a^2 - x^2)$ $a^2y - x^2y - cx^2 = 0$

    1. $a=0.2, c=0.2$
    2. $a=0.5, c=0.2$
    3. $a=0.8, c=0.2$

2.3.8

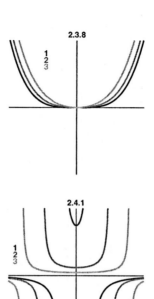

2.4.1

2.4.2

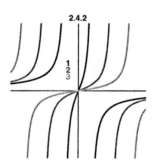

2.4.3

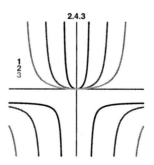

2.4.4 $y = cx^3/(a^2 - x^2)$ $a^2y - x^2y - cx^3 = 0$

        1. $a = 0.2, c = 0.2$
        2. $a = 0.5, c = 0.2$
        3. $a = 0.8, c = 0.2$

2.4.5 $y = c/[x(a^2 - x^2)]$ $a^2xy - x^3y - c = 0$

        1. $a = 0.2, c = 0.001$
        2. $a = 0.5, c = 0.001$
        3. $a = 0.8, c = 0.001$

2.4.6 $y = c/[x^2(a^2 - x^2)]$ $a^2x^2y - x^4y - c = 0$

        1. $a = 0.2, c = 0.0003$
        2. $a = 0.5, c = 0.0003$
        3. $a = 0.8, c = 0.0003$

2.4.7 $y = cx(a^2 - x^2)$ $y - a^2cx + cx^3 = 0$

        1. $a = 0.2, c = 1.0$
        2. $a = 0.5, c = 1.0$
        3. $a = 0.8, c = 1.0$

2.4.4

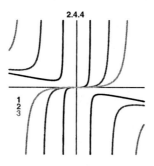

2.4.5

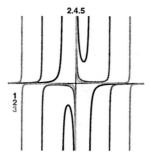

2.4.6

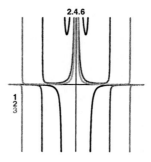

2.4.7

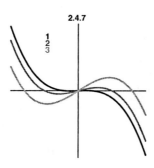

2.4.8 $y = cx^2(a^2 - x^2)$ $y - a^2cx^2 + cx^4 = 0$

         1. $a = 0.2, c = 4.0$
         2. $a = 0.5, c = 4.0$
         3. $a = 0.8, c = 4.0$

2.5   Functions with $a^3 + x^3$ and x^m

2.5.1 $y = c/(a^3 + x^3)$ $a^3y + x^3y - c = 0$

         1. $a = 0.2, c = 0.005$
         2. $a = 0.3, c = 0.005$
         3. $a = 0.4, c = 0.005$

2.5.2 $y = cx/(a^3 + x^3)$ $a^3y + x^3y - cx = 0$

         1. $a = 0.1, c = 0.01$
         2. $a = 0.3, c = 0.01$
         3. $a = 0.5, c = 0.01$

2.5.3 $y = cx^2/(a^3 + x^3)$ $a^3y + x^3y - cx^2 = 0$

         1. $a = 0.1, c = 0.1$
         2. $a = 0.3, c = 0.1$
         3. $a = 0.5, c = 0.1$

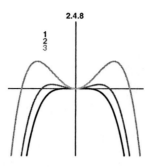

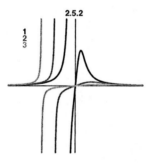

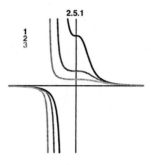

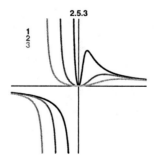

2.5.4 $y = cx^3/(a^3 + x^3)$ $a^3y + x^3y - cx^3 = 0$

        1. $a=0.1$, $c=0.2$
        2. $a=0.3$, $c=0.2$
        3. $a=0.5$, $c=0.2$

2.5.5 $y = c/[x(a^3 + x^3)]$ $a^3xy + x^4y - c = 0$

        1. $a=0.5$, $c=0.01$
        2. $a=0.7$, $c=0.01$
        3. $a=0.9$, $c=0.01$

2.5.6 $y = cx(a^3 + x^3)$ $y - a^3cx - cx^4 = 0$

        1. $a=0.5$, $c=2.0$
        2. $a=0.7$, $c=2.0$
        3. $a=0.9$, $c=2.0$

2.6 Functions with $a^3 - x^3$ and x^m

2.6.1 $y = c/(a^3 - x^3)$ $a^3y - x^3y - c = 0$

        1. $a=0.2$, $c=0.005$
        2. $a=0.3$, $c=0.005$
        3. $a=0.4$, $c=0.005$

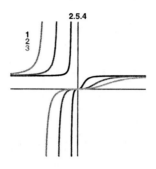

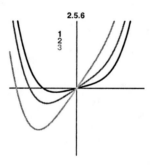

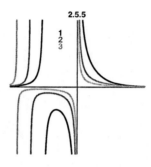

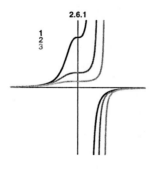

2.6.2 $y = cx/(a^3 - x^3)$ $a^3y - x^3y - cx = 0$

       1. $a = 0.1$, $c = 0.01$
       2. $a = 0.3$, $c = 0.01$
       3. $a = 0.5$, $c = 0.01$

2.6.3 $y = cx^2/(a^3 - x^3)$ $a^3y - x^3y - cx^2 = 0$

       1. $a = 0.1$, $c = 0.1$
       2. $a = 0.3$, $c = 0.1$
       3. $a = 0.5$, $c = 0.1$

2.6.4 $y = cx^3/(a^3 - x^3)$ $a^3y - x^3y - cx^3 = 0$

       1. $a = 0.1$, $c = 0.2$
       2. $a = 0.3$, $c = 0.2$
       3. $a = 0.5$, $c = 0.2$

2.6.5 $y = c/[x(a^3 - x^3)]$ $a^3xy - x^4y - c = 0$

       1. $a = 0.5$, $c = 0.01$
       2. $a = 0.7$, $c = 0.01$
       3. $a = 0.9$, $c = 0.01$

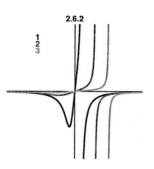

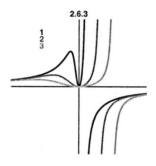

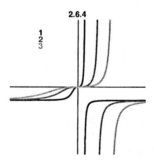

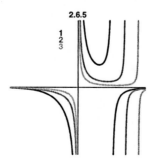

2.6.6 $y = cx(a^3 - x^3)$ $y - a^3cx + cx^4 = 0$

       1. $a=0.5$, $c=2.0$
       2. $a=0.7$, $c=2.0$
       3. $a=0.9$, $c=2.0$

2.7   Functions with $a^4 + x^4$ and x^m

2.7.1 $y = c/(a^4 + x^4)$ $a^4y + x^4y - c = 0$

       1. $a=0.3$, $c=0.007$
       2. $a=0.4$, $c=0.007$
       3. $a=0.5$, $c=0.007$

2.7.2 $y = cx/(a^4 + x^4)$ $a^4y + x^4y - cx = 0$

       1. $a=0.2$, $c=0.01$
       2. $a=0.3$, $c=0.01$
       3. $a=0.4$, $c=0.01$

2.7.3 $y = cx^2/(a^4 + x^4)$ $a^4y + x^4y - cx^2 = 0$

       1. $a=0.3$, $c=0.15$
       2. $a=0.4$, $c=0.15$
       3. $a=0.5$, $c=0.15$

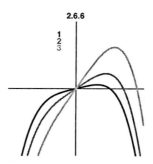

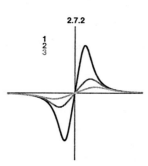

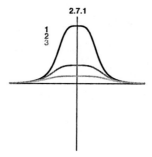

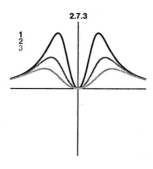

2.7.4 $\qquad y = cx^3/(a^4 + x^4) \qquad a^4y + x^4y - cx^3 = 0$

      1. $a=0.2, c=0.25$
      2. $a=0.4, c=0.25$
      3. $a=0.6, c=0.25$

2.7.5 $\qquad y = cx^4/(a^4 + x^4) \qquad a^4y + x^4y - cx^4 = 0$

      1. $a=0.2, c=1.0$
      2. $a=0.5, c=1.0$
      3. $a=0.8, c=1.0$

2.7.6 $\qquad y = cx(a^4 + x^4) \qquad y - a^4cx - cx^5 = 0$

      1. $a=0.5, c=0.5$
      2. $a=0.8, c=0.5$
      3. $a=1.1, c=0.5$

2.8 Functions with $a^4 - x^4$ and x^m

2.8.1 $\qquad y = c/(a^4 - x^4) \qquad a^4y - x^4y - c = 0$

      1. $a=0.4, c=0.01$
      2. $a=0.6, c=0.01$
      3. $a=0.8, c=0.01$

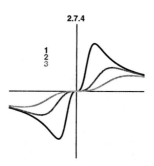

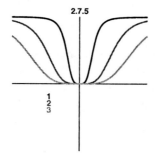

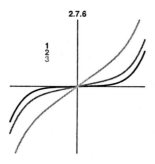

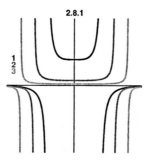

2.8.2 $y = cx/(a^4 - x^4)$ $a^4y - x^4y - cx = 0$

  1. $a=0.4, c=0.01$
  2. $a=0.6, c=0.01$
  3. $a=0.8, c=0.01$

2.8.3 $y = cx^2/(a^4 - x^4)$ $a^4y - x^4y - cx^2 = 0$

  1. $a=0.2, c=0.1$
  2. $a=0.4, c=0.1$
  3. $a=0.6, c=0.1$

2.8.4 $y = cx^3/(a^4 - x^4)$ $a^4y - x^4y - cx^3 = 0$

  1. $a=0.2, c=0.1$
  2. $a=0.4, c=0.1$
  3. $a=0.6, c=0.1$

2.8.5 $y = cx^4/(a^4 - x^4)$ $a^4y - x^4y - cx^4 = 0$

  1. $a=0.2, c=0.1$
  2. $a=0.5, c=0.1$
  3. $a=0.8, c=0.1$

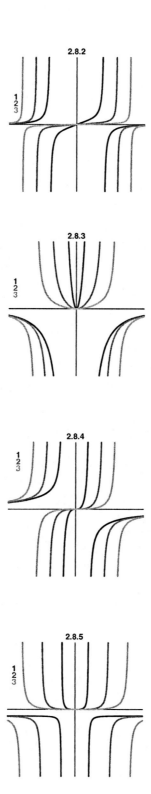

2.8.6 $y = cx(a^4 - x^4)$ $y - a^4 cx + cx^5 = 0$

              1. $a=0.4, c=1.0$
              2. $a=0.7, c=1.0$
              3. $a=1.0, c=1.0$

2.9   Functions with $(a+bx)^{1/2}$ and x^m

2.9.1 $y = c(a + bx)^{1/2}$ $y^2 - bc^2 x - ac^2 = 0$

Parabola

              1. $a=0.5, b=0.5, c=1.0$
              2. $a=0.5, b=1.0, c=1.0$
              3. $a=0.5, b=2.0, c=1.0$

2.9.2 $y = cx(a + bx)^{1/2}$ $y^2 - bc^2 x^3 - ac^2 x^2 = 0$

Special case: $c=1/(3a)$ and $b=1$ gives *Tschirnhauser's cubic* (also called *trisectrix of Catalan*)

              1. $a=0.5, b=0.5, c=1.0$
              2. $a=0.5, b=1.0, c=1.0$
              3. $a=0.5, b=2.0, c=1.0$

2.9.3 $y = cx^2(a + bx)^{1/2}$ $y^2 - bc^2 x^5 - ac^2 x^4 = 0$

              1. $a=0.5, b=0.5, c=1.0$
              2. $a=0.5, b=1.0, c=1.0$
              3. $a=0.5, b=2.0, c=1.0$

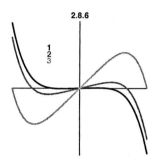

2.8.6

1
2
3

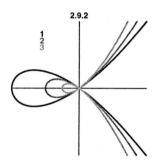

2.9.1

1
2
3

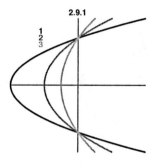

2.9.2

1
2
3

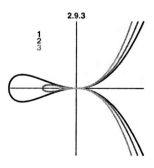

2.9.3

1
2
3

2.9.4 $y = c(a + bx)^{1/2}/x$ $x^2y^2 - c^2bx - c^2a = 0$

       1. $a=0.5, b=0.5, c=0.2$
       2. $a=0.5, b=1.0, c=0.2$
       3. $a=0.5, b=2.0, c=0.2$

2.9.5 $y = c(a + bx)^{1/2}/x^2$ $x^4y^2 - c^2bx - c^2a = 0$

       1. $a=0.5, b=0.5, c=0.1$
       2. $a=0.5, b=1.0, c=0.1$
       3. $a=0.5, b=2.0, c=0.1$

2.9.6 $y = c/(a + bx)^{1/2}$ $ay^2 + bxy^2 - c^2 = 0$

       1. $a=1.0, b=0.5, c=0.5$
       2. $a=1.0, b=1.0, c=0.5$
       3. $a=1.0, b=2.0, c=0.5$

2.9.7 $y = cx/(a + bx)^{1/2}$ $ay^2 + bxy^2 - c^2x^2 = 0$

       1. $a=1.0, b=1.0, c=1.0$
       2. $a=1.0, b=2.0, c=1.0$
       3. $a=1.0, b=4.0, c=1.0$

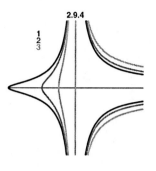

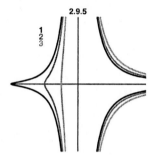

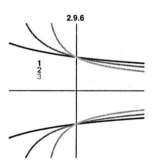

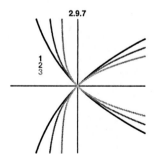

2.9.8 $y = cx^2/(a + bx)^{1/2}$ $ay^2 + bxy^2 - c^2x^4 = 0$

       1. $a=1.0, b=1.0, c=1.0$
       2. $a=1.0, b=2.0, c=1.0$
       3. $a=1.0, b=4.0, c=1.0$

2.9.9 $y = c/[x(a + bx)^{1/2}]$ $ax^2y^2 + bx^3y^2 - c^2 = 0$

       1. $a=1.0, b=0.8, c=0.2$
       2. $a=1.0, b=1.0, c=0.2$
       3. $a=1.0, b=1.2, c=0.2$

2.9.10 $y = c/[x^2(a + bx)^{1/2}]$ $ax^4y^2 + bx^5y^2 - c^2 = 0$

       1. $a=1.0, b=0.8, c=0.1$
       2. $a=1.0, b=1.0, c=0.1$
       3. $a=1.0, b=1.2, c=0.1$

2.9.11 $y = c[x(a + bx)]^{1/2}$ $y^2 - ac^2x - bc^2x^2 = 0$

       1. $a=2.0, b=-2.0, c=1.0$
       2. $a=2.0, b=-3.0, c=1.0$
       3. $a=2.0, b=-4.0, c=1.0$

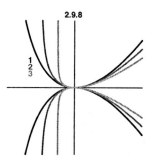

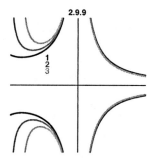

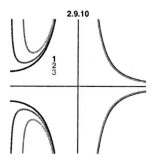

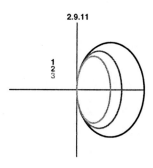

4. $a=2.0, b=3.0, c=0.4$
5. $a=2.0, b=5.0, c=0.4$
6. $a=2.0, b=7.0, c=0.4$

2.9.12 $y = c[x^3(a + bx)]^{1/2}$ $y^2 - ac^2x^3 - bc^2x^4 = 0$

Special case: $b < 0$ gives *piriform*

1. $a=1.0, b=-1.0, c=2.0$
2. $a=1.0, b=-1.5, c=2.0$
3. $a=1.0, b=-2.0, c=2.0$

4. $a=2.0, b=3.0, c=0.4$
5. $a=2.0, b=5.0, c=0.4$
6. $a=2.0, b=7.0, c=0.4$

2.9.13 $y = c[(a + bx)/x]^{1/2}$ $xy^2 - c^2bx - c^2a = 0$

1. $a=2.0, b=-2.0, c=0.1$
2. $a=2.0, b=-3.0, c=0.1$
3. $a=2.0, b=-6.0, c=0.1$

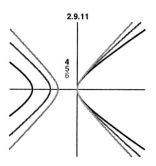

2.9.11

4
5
6

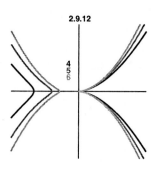

2.9.12

1
2
3

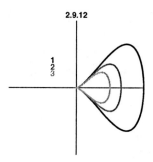

2.9.12

4
5
6

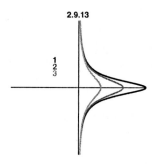

2.9.13

1
2
3

    4. $a=2.0, b=4.0, c=0.1$
    5. $a=2.0, b=6.0, c=0.1$
    6. $a=2.0, b=8.0, c=0.1$

2.9.14 $y = c[(a + bx)/x^3]^{1/2}$ $x^3y - c^2bx - c^2a = 0$

    1. $a=2.0, b=-2.0, c=0.1$
    2. $a=2.0, b=-3.0, c=0.1$
    3. $a=2.0, b=-6.0, c=0.1$

    4. $a=2.0, b=4.0, c=0.1$
    5. $a=2.0, b=6.0, c=0.1$
    6. $a=2.0, b=8.0, c=0.1$

2.9.15 $y = c[x/(a + bx)]^{1/2}$ $ay^2 + bxy^2 - c^2x = 0$

    1. $a=4.0, b=-2.0, c=1.0$
    2. $a=4.0, b=-4.0, c=1.0$
    3. $a=4.0, b=-8.0, c=1.0$

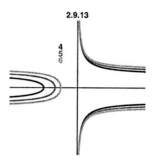

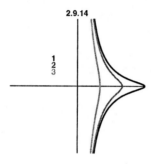

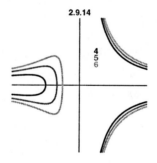

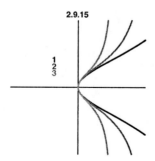

$\qquad$ 4. $a=1.0$, $b=3.0$, $c=1.0$
$\qquad$ 5. $a=1.0$, $b=5.0$, $c=1.0$
$\qquad$ 6. $a=1.0$, $b=8.0$, $c=1.0$

2.9.16 $\qquad$ $y = c[x^3/(a+bx)]^{1/2}$ $\qquad$ $ay^2 + bxy^2 - c^2x^3 = 0$

Special case: $b = -a$ gives *cissoid of Diocles*

$\qquad$ 1. $a=1.0$, $b=-0.5$, $c=1.0$
$\qquad$ 2. $a=1.0$, $b=-1.0$, $c=1.0$
$\qquad$ 3. $a=1.0$, $b=-2.0$, $c=1.0$

$\qquad$ 4. $a=1.0$, $b=3.0$, $c=1.0$
$\qquad$ 5. $a=1.0$, $b=4.0$, $c=1.0$
$\qquad$ 6. $a=1.0$, $b=8.0$, $c=1.0$

2.10 Functions with $(a^2 - x^2)^{1/2}$ and x^m

2.10.1 $\qquad$ $y = c(a^2 - x^2)^{1/2}$ $\qquad$ $y^2 + c^2x^2 - a^2c^2 = 0$

Ellipse ($c=1$ gives *circle*)

$\qquad$ 1. $a=0.5$, $c=1.0$
$\qquad$ 2. $a=0.5$, $c=1.5$
$\qquad$ 3. $a=0.5$, $c=2.0$

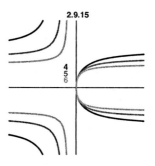

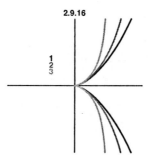

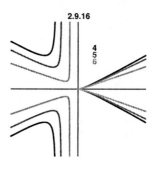

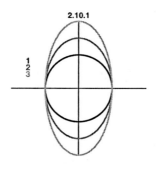

2.10.2 $y = cx(a^2 - x^2)^{1/2}$ $y^2 - c^2a^2x^2 + c^2x^4 = 0$

Eight curve (also called *lemniscate of Gerono*)

        1. $a=0.6, c=1.0$
        2. $a=0.8, c=1.0$
        3. $a=1.0, c=1.0$

2.10.3 $y = cx^2(a^2 - x^2)^{1/2}$ $y^2 - c^2a^2x^4 + c^2x^6 = 0$

        1. $a=0.6, c=2.0$
        2. $a=0.8, c=2.0$
        3. $a=1.0, c=2.0$

2.10.4 $y = c(a^2 - x^2)^{1/2}/x$ $x^2y^2 + c^2x^2 - c^2a^2 = 0$

        1. $a=0.50, c=0.1$
        2. $a=0.75, c=0.1$
        3. $a=1.00, c=0.1$

2.10.5 $y = c(a^2 - x^2)^{1/2}/x^2$ $x^4y^2 + c^2x^2 - c^2a^2 = 0$

        1. $a=0.50, c=0.1$
        2. $a=0.75, c=0.1$
        3. $a=1.00, c=0.1$

2.10.2

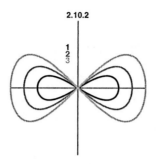

2.10.3

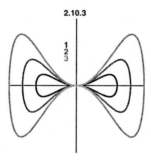

2.10.4

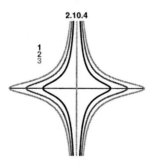

2.10.5

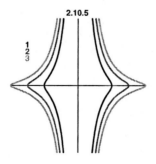

2.10.6 $y = c/(a^2 - x^2)^{1/2}$ $a^2 y^2 - x^2 y^2 - c^2 = 0$

        1. $a = 0.50$, $c = 0.1$
        2. $a = 0.75$, $c = 0.1$
        3. $a = 1.00$, $c = 0.1$

2.10.7 $y = c/[x(a^2 - x^2)^{1/2}]$ $a^2 x^2 y^2 - x^4 y^2 - c^2 = 0$

        1. $a = 0.50$, $c = 0.1$
        2. $a = 0.75$, $c = 0.1$
        3. $a = 1.00$, $c = 0.1$

2.10.8 $y = cx/(a^2 - x^2)^{1/2}$ $a^2 y^2 - x^2 y^2 - c^2 x^2 = 0$

Bullet-nose curve

        1. $a = 0.50$, $c = 0.4$
        2. $a = 0.75$, $c = 0.4$
        3. $a = 1.00$, $c = 0.4$

2.10.9 $y = cx^2/(a^2 - x^2)^{1/2}$ $a^2 y^2 - x^2 y^2 - c^2 x^4 = 0$

        1. $a = 0.50$, $c = 0.4$
        2. $a = 0.75$, $c = 0.4$
        3. $a = 1.00$, $c = 0.4$

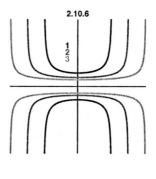

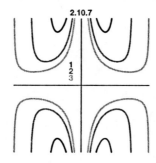

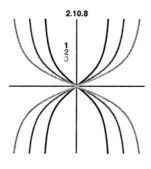

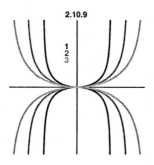

2.11   Functions with $(x^2 - a^2)^{1/2}$ and x^m

2.11.1 $y = c(x^2 - a^2)^{1/2}$ $y^2 - c^2 x^2 + a^2 c^2 = 0$

Hyperbola

> 1. $a=0.1, c=1.00$
> 2. $a=0.3, c=1.00$
> 3. $a=0.5, c=1.00$

2.11.2 $y = cx(x^2 - a^2)^{1/2}$ $y^2 - c^2 x^4 + c^2 a^2 x^2 = 0$

Kampyle of Eudoxus

> 1. $a=0.1, c=1.0$
> 2. $a=0.4, c=1.0$
> 3. $a=0.7, c=1.0$

2.11.3 $y = cx^2(x^2 - a^2)^{1/2}$ $y^2 - c^2 x^6 + c^2 a^2 x^4 = 0$

> 1. $a=0.1, c=1.0$
> 2. $a=0.4, c=1.0$
> 3. $a=0.7, c=1.0$

2.11.4 $y = c(x^2 - a^2)^{1/2}/x$ $x^2 y^2 - c^2 x^2 + c^2 a^2 = 0$

> 1. $a=0.1, c=1.0$
> 2. $a=0.4, c=1.0$
> 3. $a=0.7, c=1.0$

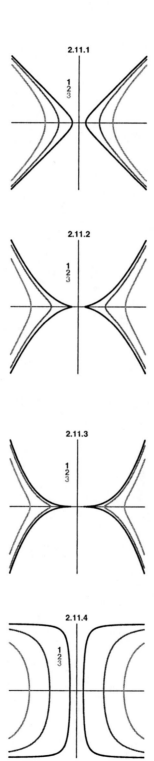

2.11.5 $y = c(x^2 - a^2)^{1/2}/x^2$ $x^4y^2 - c^2x^2 + c^2a^2 = 0$

        1. $a=0.1, c=0.2$
        2. $a=0.2, c=0.2$
        3. $a=0.3, c=0.2$

2.11.6 $y = c/(x^2 - a^2)^{1/2}$ $x^2y^2 - a^2y^2 - c^2 = 0$

        1. $a=0.1, c=0.1$
        2. $a=0.3, c=0.1$
        3. $a=0.5, c=0.1$

2.11.7 $y = c/[x(x^2 - a^2)^{1/2}]$ $x^4y^2 - a^2x^2y^2 - c^2 = 0$

        1. $a=0.1, c=0.02$
        2. $a=0.3, c=0.02$
        3. $a=0.5, c=0.02$

2.11.8 $y = cx/(x^2 - a^2)^{1/2}$ $x^2y^2 - a^2y^2 - c^2x^2 = 0$

Cross curve

        1. $a=0.1, c=0.1$
        2. $a=0.3, c=0.1$
        3. $a=0.5, c=0.1$

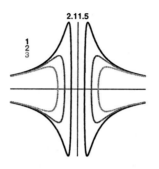

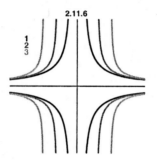

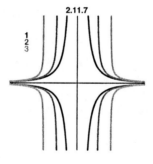

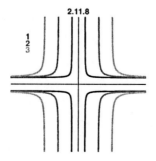

2.11.9 $y = cx^2/(x^2 - a^2)^{1/2}$ $x^2y^2 - a^2y^2 - c^2x^4 = 0$

    1. $a=0.1, c=0.5$
    2. $a=0.3, c=0.5$
    3. $a=0.5, c=0.5$

2.12   Functions with $(a^2 + x^2)^{1/2}$ and x^m

2.12.1 $y = c(a^2 + x^2)^{1/2}$ $y^2 - c^2a^2 - c^2x^2 = 0$

    1. $a=0.1, c=1.0$
    2. $a=0.3, c=1.0$
    3. $a=0.5, c=1.0$

2.12.2 $y = cx(a^2 + x^2)^{1/2}$ $y^2 - c^2a^2x^2 - c^2x^4 = 0$

    1. $a=1.0, c=0.5$
    2. $a=2.0, c=0.5$
    3. $a=4.0, c=0.5$

2.12.3 $y = cx^2(a^2 + x^2)^{1/2}$ $y^2 - c^2a^2x^4 - c^2x^6 = 0$

    1. $a=1.0, c=0.5$
    2. $a=2.0, c=0.5$
    3. $a=4.0, c=0.5$

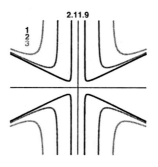

2.11.9

1
2
3

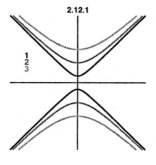

2.12.1

1
2
3

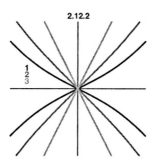

2.12.2

1
2
3

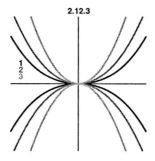

2.12.3

1
2
3

2.12.4 $y = c(a^2 + x^2)^{1/2}/x$ $x^2y^2 - c^2x^2 - c^2a^2 = 0$

  1. $a=0.2, c=0.2$
  2. $a=0.5, c=0.2$
  3. $a=0.8, c=0.2$

2.12.5 $y = c(a^2 + x^2)^{1/2}/x^2$ $x^4y^2 - c^2x^2 - c^2a^2 = 0$

  1. $a=0.2, c=0.2$
  2. $a=0.5, c=0.2$
  3. $a=0.8, c=0.2$

2.12.6 $y = c/(a^2 + x^2)^{1/2}$ $a^2y^2 + x^2y^2 - c^2 = 0$

  1. $a=0.2, c=0.2$
  2. $a=0.4, c=0.2$
  3. $a=0.8, c=0.2$

2.12.7 $y = c/[x(a^2 + x^2)^{1/2}]$ $a^2x^2y^2 + x^4y^2 - c^2 = 0$

  1. $a=1.0, c=0.5$
  2. $a=2.0, c=0.5$
  3. $a=4.0, c=0.5$

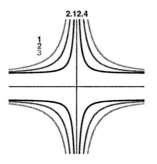

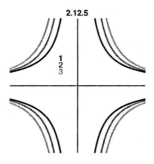

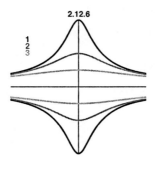

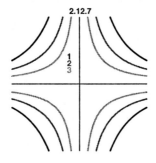

2.12.8 $y = cx/(a^2 + x^2)^{1/2}$ $a^2y^2 + x^2y^2 - c^2x^2 = 0$

        1. $a=0.5, c=1.0$
        2. $a=1.0, c=1.0$
        3. $a=2.0, c=1.0$

2.12.9 $y = cx^2/(a^2 + x^2)^{1/2}$ $a^2y^2 + x^2y^2 - c^2x^4 = 0$

        1. $a=0.1, c=1.0$
        2. $a=0.5, c=1.0$
        3. $a=1.0, c=1.0$

2.13 Miscellaneous Functions

2.13.1 $y = c(a + x)/(b - x)$ $by - xy - cx - ac = 0$

        1. $a=0.5, b=0.1, c=0.1$
        2. $a=0.5, b=0.3, c=0.1$
        3. $a=0.5, b=0.5, c=0.1$

        4. $a=0.5, b=0.3, c=0.1$
        5. $a=1.0, b=0.3, c=0.1$
        6. $a=1.5, b=0.3, c=0.1$

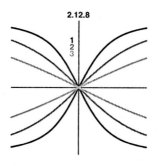

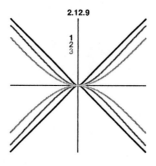

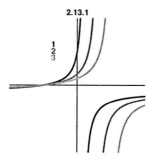

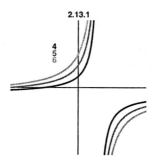

2.13.2 $y = c[(a + x)/(b - x)]^{1/2}$ $by^2 - xy^2 - c^2x - ac^2 = 0$

        1. $a=1.0,\ b=0.1,\ c=0.2$
        2. $a=1.0,\ b=0.3,\ c=0.2$
        3. $a=1.0,\ b=0.5,\ c=0.2$

        4. $a=0.0,\ b=0.8,\ c=0.2$
        5. $a=0.5,\ b=0.8,\ c=0.2$
        6. $a=1.0,\ b=0.8,\ c=0.2$

2.13.3 $y = cx[(a + x)/(b - x)]^{1/2}$ $by^2 - xy^2 - ac^2x^2 - c^2x^3 = 0$

Special cases: $a=b$ gives *right strophoid*; $a=3b$ gives *trisectrix of Maclaurin*

        1. $a=1.0,\ b=-0.3,\ c=0.4$
        2. $a=1.0,\ b=0.0,\ c=0.4$
        3. $a=1.0,\ b=0.3,\ c=0.4$

        4. $a=0.0,\ b=0.8,\ c=0.4$
        5. $a=0.5,\ b=0.8,\ c=0.4$
        6. $a=1.0,\ b=0.8,\ c=0.4$

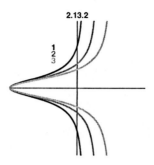

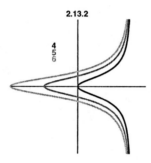

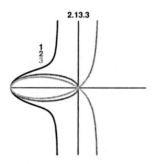

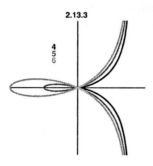

2.13.4 $y = (c/x)[(a + x)/(b - x)]^{1/2}$ $bx^2y^2 - x^3y^2 - c^2x - ac^2 = 0$

>   1. $a=1.0, b=0.4, c=0.05$
>   2. $a=1.0, b=0.6, c=0.05$
>   3. $a=1.0, b=0.8, c=0.05$

>   4. $a=-0.5, b=0.8, c=0.05$
>   5. $a=0.0, b=0.8, c=0.05$
>   6. $a=0.5, b=0.8, c=0.05$

$$y = cx\{[a^2/(x - b)^2] - 1\}^{1/2}$$

2.13.5

$$b^2y^2 - 2bxy^2 + x^2y^2 + b^2x^2 - a^2x^2 - 2bx^3 + x^4 = 0$$

Conchoid of Nicomedes (also called *cochloid*)

>   1. $a=0.5, b=0.10, c=1.0$
>   2. $a=0.5, b=0.25, c=1.0$
>   3. $a=0.5, b=0.50, c=1.0$

>   4. $a=0.5, b=0.50, c=1.0$
>   5. $a=1.0, b=0.50, c=1.0$
>   6. $a=1.5, b=0.50, c=1.0$

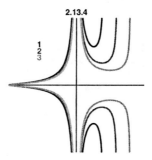

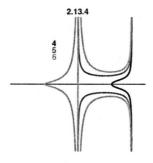

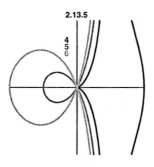

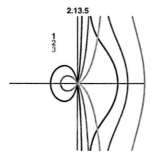

2.13.6 $y = c(a^2 + x^2)/(b^2 - x^2)$ $b^2y - x^2y - cx^2 - a^2c = 0$

       1. $a=0.5, b=0.4, c=0.1$
       2. $a=0.5, b=0.6, c=0.1$
       3. $a=0.5, b=0.8, c=0.1$

       4. $a=0.5, b=0.5, c=0.1$
       5. $a=0.7, b=0.5, c=0.1$
       6. $a=0.9, b=0.5, c=0.1$

2.13.7 $y = c[x - (x^2 - a^2)^{1/2}]$ $y^2 - 2xy + a^2 = 0$

       1. $a=0.1, c=1.0$
       2. $a=0.3, c=1.0$
       3. $a=0.5, c=1.0$

2.13.8 $y = c[x - (a^2 + x^2)^{1/2}]$ $y^2 - 2xy - a^2 = 0$

       1. $a=0.1, c=1.0$
       2. $a=0.3, c=1.0$
       3. $a=0.5, c=1.0$

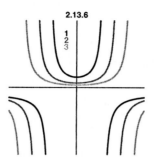

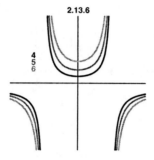

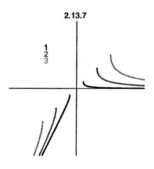

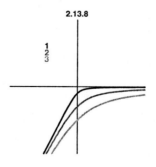

2.13.9 $\qquad y = c(a^2 - x^2)[2a \pm (a^2 - x^2)^{1/2}]/(3a^2 + x^2) \qquad y^2(a^2 - x^2) - (x^2 + 2ay - a^2)^2 = 0$

Bicorn

        1. $a = 0.50$, $c = 1.0$
        2. $a = 0.75$, $c = 1.0$
        3. $a = 1.00$, $c = 1.0$

2.13.10 $\qquad y = c(1 - |x/a|^{n/m})^{m/n} \qquad y^{n/m} + (c|x/a|)^{n/m} - c^{n/m} = 0$

Hyperellipse (also called *Lame curve*) for $n/m > 2$; *hypoellipse* for $n/m < 2$

        1. $a = 0.5$, $n/m = 1/3$, $c = 1.0$
        2. $a = 0.5$, $n/m = 2/3$, $c = 1.0$
        3. $a = 0.5$, $n/m = 1$, $c = 1.0$

        4. $a = 0.5$, $n/m = 2$, $c = 1.0$
        5. $a = 0.5$, $n/m = 5$, $c = 1.0$
        6. $a = 0.5$, $n/m = 20$, $c = 1.0$

2.13.11 $\qquad y = c(1 + |x/a|^{n/m})^{m/n} \qquad y^{n/m} - (c|x/a|)^{n/m} - c^{m/n} = 0$

        1. $a = 0.2$, $n/m = 1/3$, $c = 0.2$
        2. $a = 0.2$, $n/m = 2/3$, $c = 0.2$
        3. $a = 0.2$, $n/m = 1$, $c = 0.2$

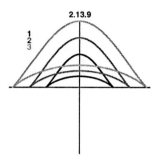

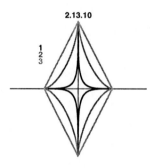

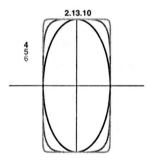

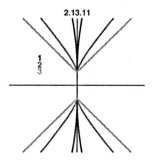

$$4.\ a=0.2,\ n/m=2,\ c=0.2$$
$$5.\ a=0.2,\ n/m=5,\ c=0.2$$
$$6.\ a=0.2,\ n/m=20,\ c=0.2$$

2.13.12 $y = cx^{ax}$ $y - cx^{ax} = 0$

   1. $a=2.0,\ c=1.0$
   2. $a=5.0,\ c=1.0$
   3. $a=8.0,\ c=1.0$

2.13.13 $y = c(1-x^2)^{ax}$ $y - c(1-x^2)^{ax} = 0$

   1. $a=1.0,\ c=0.5$
   2. $a=2.0,\ c=0.5$
   3. $a=4.0,\ c=0.5$

2.13.14 $y = c(1-x^2)^{1/ax}$ $y - c(1-x^2)^{1/ax} = 0$

   1. $a=1.0,\ c=0.5$
   2. $a=2.0,\ c=0.5$
   3. $a=4.0,\ c=0.5$

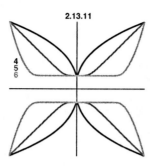

2.13.11

4
5
6

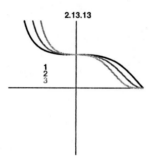

2.13.12

1
2
3

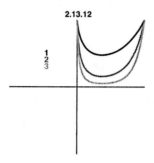

2.13.13

1
2
3

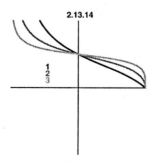

2.13.14

1
2
3

2.13.15 $y = cx(1 + a^2x^2)^2 \qquad y - cx(1 + a^2x^2)^2 = 0$

       1. $a = 1.0, c = 0.1$
       2. $a = 2.0, c = 0.1$
       3. $a = 4.0, c = 0.1$

2.13.16 $y = cx^2(1 + a^2x^2)^2 \qquad y - cx^2(1 + a^2x^2)^2 = 0$

       1. $a = 1.0, c = 0.1$
       2. $a = 2.0, c = 0.1$
       3. $a = 4.0, c = 0.1$

2.13.17 $x^n - \binom{n}{2}x^{n-2}y^2 + \binom{n}{4}x^{n-4}y^4 - \cdots = a^n$

N-roll mill (implicit form only)

       a. $a = 0.5, n = 2$

       b. $a = 0.5, n = 3$

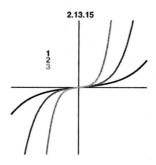

2.13.15

1
2
3

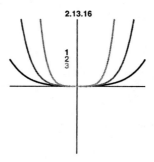

2.13.16

1
2
3

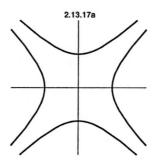

2.13.17a

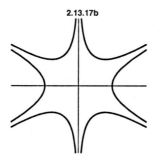

2.13.17b

2.14 Functions Expressible in Polar Coordinates

2.14.1 $r = c\theta^{n/m}$ $(x^2 + y^2)^{m/(2n)} - c^{m/n}\arctan(y/x) = 0$

         a. $c = 0.04$, $m = 1$; $0 < \theta < 8\pi$ (*Archimedes' spiral*)

         b. $c = 0.20$, $m = 2$; $0 < \theta < 8\pi$ (*Fermat's spiral*)

2.14.2 $r = c/\theta^{n/m}$ $(x^2 + y^2)^{m/(2n)}\arctan(y/x) - c^{m/n} = 0$

         a. $c = 1.0$, $m = 1$; $0.50 < \theta < 6\pi$ (*hyperbolic spiral*)

         b. $c = 0.5$, $m = 2$; $0.25 < \theta < 6\pi$ (*lituus*)

2.14.1a

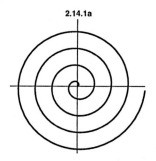

2.14.1b

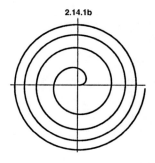

2.14.2a

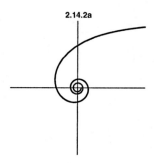

2.14.2b

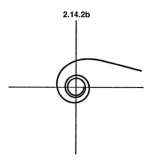

2.14.3 $r = a + b\theta^{n/m}$ $[(x^2 + y^2)^{1/2} - a]^{m/n} - b^{m/n}\arctan(y/x) = 0$

Special case: $n/m = 1/2$ gives *parabolic spiral*

a. $a = 0.1$, $b = 0.2$, $n = 1$, $m = 2$; $0 < \theta < 6\pi$

b. $a = 0.1$, $b = 0.2$, $n = 1$, $m = 4$; $0 < \theta < 8\pi$

c. $a = 0.4$, $b = 0.1$, $n = 1$, $m = 2$; $0 < \theta < 8\pi$

d. $a = 0.4$, $b = 0.1$, $n = 1$, $m = 4$; $0 < \theta < 8\pi$

2.14.3a

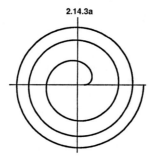

2.14.3b

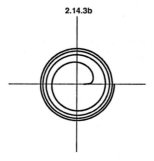

2.14.3c

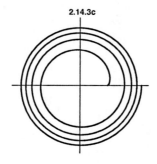

2.14.3d

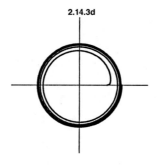

2.14.4 $r = a + b/\theta^{n/m}$ $[(x^2 + y^2)^{1/2} - a]^{m/n} \arctan(y/x) - b^{m/n} = 0$

a. $a = 0.25$, $b = 0.75$, $n = 1$, $m = 2$; $0 < \theta < 8\pi$

b. $a = 0.25$, $b = 0.75$, $n = 1$, $m = 4$; $0 < \theta < 8\pi$

c. $a = 0.25$, $b = 0.25$, $n = 1$, $m = 2$; $0 < \theta < 8\pi$

d. $a = 0.25$, $b = 0.25$, $n = 1$, $m = 4$; $0 < \theta < 8\pi$

2.14.4a

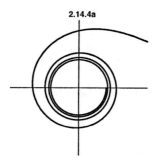

2.14.4b

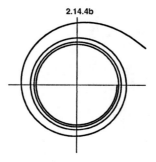

2.14.4c

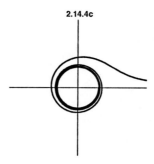

2.14.4d

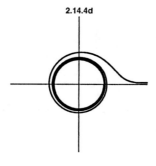

2.14.5 $r = c(a^2 + \theta^2)^{1/2}$ $(x^2 + y^2)^{1/2} - c\{a^2 + [\arctan(y/x)]^2\}^{1/2} = 0$

Special case: $a = 1$ gives *involute of a circle*

a. $a = 1.0$, $c = 0.04$; $0 < \theta < 6\pi$

b. $a = 4.0$, $c = 0.04$; $0 < \theta < 6\pi$

2.15 Functions Expressed Parametrically

2.15.1 $x = c(8at^3 + 24t^5)$; $y = c(-6at^2 - 15t^4)$

Butterfly catastrophe

a. $a = -4.0$, $c = 0.03$; $-3 < t < 3$

b. $a = -7.0$, $c = 0.02$; $-4 < t < 4$

2.14.5a

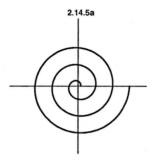

2.14.5b

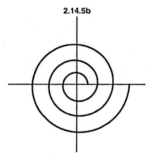

2.15.1a

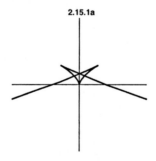

2.15.1b

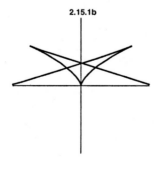

2.15.2 $x = c(-2at - 4t^3); \qquad y = c(at^2 + 3t^4)$

Swallowtail catastrophe

a. $a = -1.0, c = 0.5; \; -1 < t < 1$

b. $a = -2.0, c = 0.5; \; -2 < t < 2$

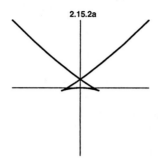

2.15.2a

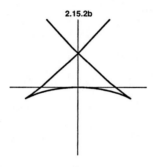

2.15.2b

3

Transcendental Functions

This chapter treats the transcendental functions: trigonometric, logarithmic, and exponential. The equations presented in this chapter can mostly be found in standard texts and in tables of integrals. Traditional or accepted names for certain curves are included wherever appropriate. The last two sections of this chapter comprise curves that are more easily expressed in polar coordinates or parametrically.

3.1 Functions with sin"(*ax*) and cos"'(*bx*) (*n, m* integers)

3.1.1 $y = \sin(2\pi x)$

3.1.2 $y = \cos(2\pi x)$

3.1.3 $y = 0.2 \tan(2\pi x)$

3.1.4 $y = 0.2 \cot(2\pi x)$

3.1.1

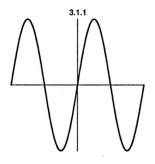

3.1.2

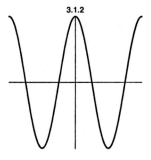

3.1.3

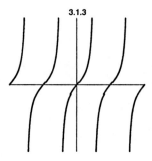

3.1.4

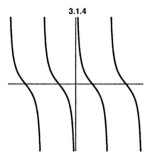

3.1.5 $y = 0.25 \csc(2\pi x)$

3.1.6 $y = 0.25 \sec(2\pi x)$

3.1.7 $y = \sin^2(2\pi x)$

3.1.8 $y = \cos^2(2\pi x)$

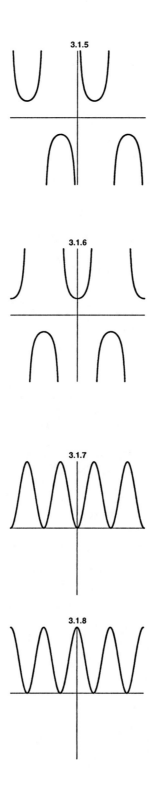

3.1.9 $y = \sin(2\pi ax)\sin(2\pi bx)$

Modulated sine wave

         1. $a=0.5$, $b=1.0$
         2. $a=0.5$, $b=1.5$

         3. $a=0.5$, $b=2.0$
         4. $a=0.5$, $b=2.5$

3.1.10 $y = \cos(2\pi ax)\cos(2\pi bx)$

         1. $a=0.5$, $b=1.0$
         2. $a=0.5$, $b=1.5$

         3. $a=0.5$, $b=2.0$
         4. $a=0.5$, $b=2.5$

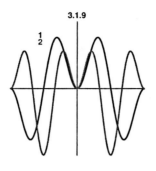

3.1.9

$\frac{1}{2}$

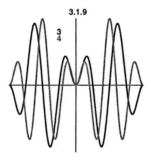

3.1.9

$\frac{3}{4}$

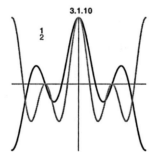

3.1.10

$\frac{1}{2}$

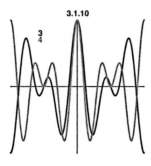

3.1.10

$\frac{3}{4}$

3.1.11 $y = \sin(2\pi ax)\cos(2\pi bx)$

        1. $a=0.5, b=1.0$
        2. $a=0.5, b=1.5$

        3. $a=0.5, b=2.0$
        4. $a=0.5, b=2.5$

3.1.12 $y = 2.0\sin(2\pi x)\cos^2(2\pi x)$

3.1.13 $y = 2.0\cos(2\pi x)\sin^2(2\pi x)$

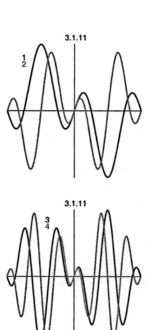

3.1.11

$\frac{1}{2}$

3.1.11

$\frac{3}{4}$

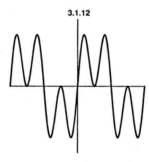

3.1.12

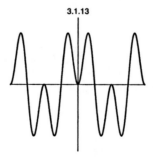

3.1.13

3.1.14 $y = 0.25 \sin(2\pi x)/\cos^2(2\pi x)$

3.1.15 $y = 0.25 \sin^2(2\pi x)/\cos(2\pi x)$

3.1.16 $y = 0.25 \cos(2\pi x)/\sin^2(2\pi x)$

3.1.17 $y = 0.25 \cos^2(2\pi x)/\sin(2\pi x)$

3.1.14

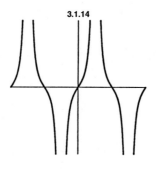

3.1.15

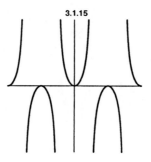

3.1.16

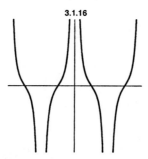

3.1.17

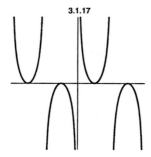

3.2   Functions with $1 \pm a \sin''(cx)$ and $1 \pm b \cos'''(cx)$

3.2.1 $y = 0.5/[1 + \cos(2\pi x)]$

3.2.2 $y = 0.5/[1 - \cos(2\pi x)]$

3.2.3 $y = 0.5 \sin(2\pi x)/[1 + \cos(2\pi x)]$

3.2.4 $y = 0.5 \sin(2\pi x)/[1 - \cos(2\pi x)]$

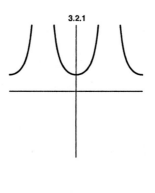

3.2.1

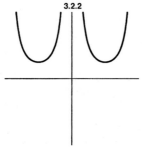

3.2.2

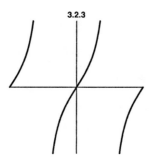

3.2.3

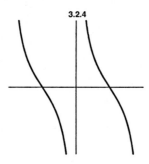

3.2.4

3.2.5 $y = 0.5 \cos(2\pi x)/[1 + \cos(2\pi x)]$

3.2.6 $y = 0.5 \cos(2\pi x)/[1 - \cos(2\pi x)]$

3.2.7 $y = 0.5/[1 + \cos(2\pi x)]^{1/2}$

3.2.8 $y = 0.5/[1 - \cos(2\pi x)]^{1/2}$

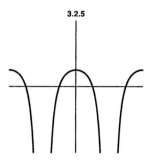

3.2.5

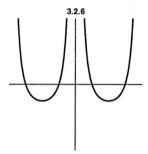

3.2.6

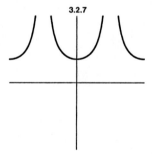

3.2.7

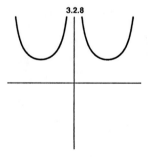

3.2.8

3.2.9 $y = 0.5/[a^2 + b^2\cos^2(2\pi x)]$

> 1. $a=0.5, b=1.0$
> 2. $a=1.0, b=1.0$
> 3. $a=2.0, b=1.0$

3.2.10 $y = 0.2/[a^2 - b^2\cos^2(2\pi x)]$

> 1. $a=0.5, b=1.0$
> 2. $a=1.0, b=1.0$
> 3. $a=1.5, b=1.0$

3.2.11 $y = [\sin(2\pi x)]/[1 + \cos^2(2\pi x)]$

3.2.12 $y = [\cos(2\pi x)]/[1 + \sin^2(2\pi x)]$

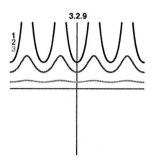

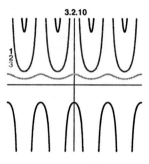

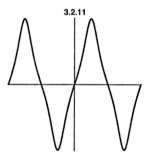

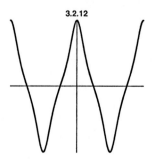

3.2.13 $y = 2.0[\sin(2\pi x)]/[1 + \sin^2(2\pi x)]$

3.2.14 $y = 2.0[\cos(2\pi x)]/[1 + \cos^2(2\pi x)]$

3.2.15 $y = [\sin^2(2\pi x)]/[1 + \cos^2(2\pi x)]$

3.2.16 $y = [\cos^2(2\pi x)]/[1 + \sin^2(2\pi x)]$

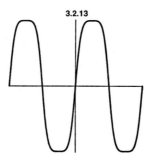

3.2.13

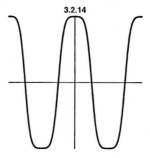

3.2.14

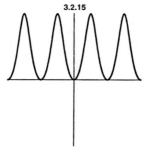

3.2.15

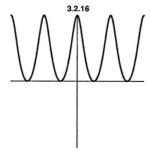

3.2.16

3.2.17 $y = 2.0[\sin^2(2\pi x)]/[1 + \sin^2(2\pi x)]$

3.2.18 $y = 2.0[\cos^2(2\pi x)]/[1 + \cos^2(2\pi x)]$

3.3   Functions with $a\ \sin''(cx) + b\ \cos'''(cx)$

3.3.1 $y = a\cos(2\pi x) + b\sin(2\pi x)$

      1. $a=0.6,\ b=0.4$
      2. $a=0.6,\ b=0.6$
      3. $a=0.6,\ b=0.8$

3.3.2 $y = 1/[a\cos(2\pi x) + b\sin(2\pi x)]$

      1. $a=1.0,\ b=1.0$
      2. $a=1.0,\ b=2.0$
      3. $a=1.0,\ b=4.0$

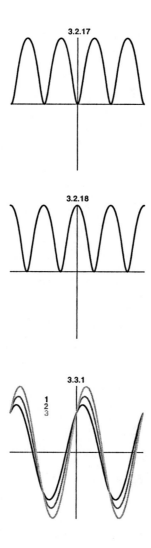

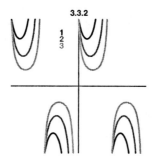

3.3.3 $y = a^2\cos^2(2\pi x) + b^2\sin^2(2\pi x)$

        1. $a=1.0, b=0.4$
        2. $a=1.0, b=0.6$
        3. $a=1.0, b=0.8$

3.3.4 $y = 1/[a^2\cos^2(2\pi x) + b^2\sin^2(2\pi x)]$

        1. $a=1.0, b=1.5$
        2. $a=1.0, b=2.0$
        3. $a=1.0, b=4.0$

3.3.5 $y = \sin(2\pi x)/[a\cos(2\pi x) + b\sin(2\pi x)]$

        1. $a=1.0, b=2.0$
        2. $a=1.0, b=4.0$
        3. $a=1.0, b=6.0$

3.3.6 $y = \cos(2\pi x)/[a\cos(2\pi x) + b\sin(2\pi x)]$

        1. $a=1.0, b=2.0$
        2. $a=1.0, b=4.0$
        3. $a=1.0, b=6.0$

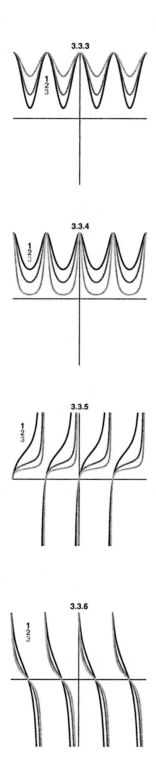

3.4 Functions of More Complicated Arguments

3.4.1 $y = \sin(a\pi/x)$

$a = 1.0$

3.4.2 $y = \cos(a\pi/x)$

$a = 1.0$

3.4.3 $y = \sin(a\pi|x|^{n/m})$

        a. $a = 4.0,\ n/m = \frac{1}{2}$

        b. $a = 4.0,\ n/m = 2$

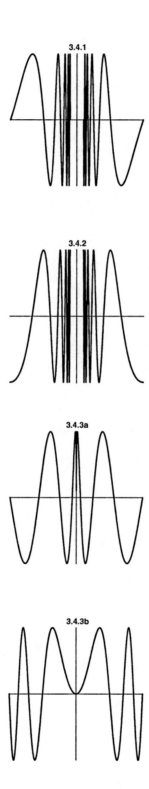

3.4.4 $y = \cos(a\pi |x|^{n/m})$

              a. $a=4.0$, $n/m=\frac{1}{2}$

              b. $a=4.0$, $n/m=2$

3.4.5 $y = \sin[\pi \cos(a\pi x)/2]$

$a=2.0$

3.4.6 $y = \sin[\pi \sin(a\pi x)/2]$

$a=2.0$

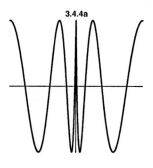

3.4.4a

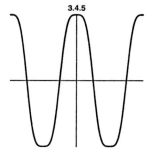

3.4.4b

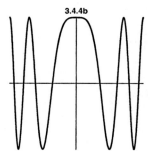

3.4.5

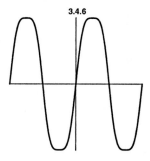

3.4.6

3.4.7 $y = \cos[\pi \sin(a\pi x)/2]$

$a = 2.0$

3.4.8 $y = \cos[\pi \cos(a\pi x)/2]$

$a = 2.0$

3.5 Inverse Trigonometric Functions

3.5.1 $y = (1/\pi)\arcsin(x)$

3.5.2 $y = (1/\pi)\arccos(x)$

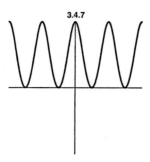

3.4.7

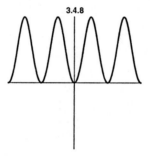

3.4.8

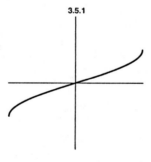

3.5.1

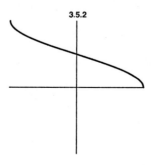

3.5.2

3.5.3 $y = (1/\pi)\arctan(10x)$

3.5.4 $y = (1/\pi)\text{arccot}(10x)$

3.5.5 $y = (1/\pi)\text{arcsec}(10x)$

3.5.6 $y = (1/\pi)\text{arccsc}(10x)$

3.5.3

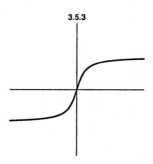

3.5.4

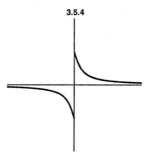

3.5.5

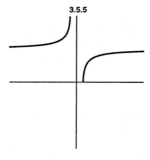

3.5.6

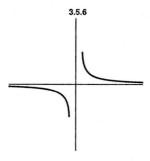

3.6   Logarithmic Functions

3.6.1 $y = 0.25 \ln(10x)$

3.6.2 $y = 0.25 \ln(1/10x)$

3.6.3 $y = 0.25/\ln(10x)$

3.6.4 $y = 0.5 \ln[(x + a)/(x - a)]$

1. $a = 0.1$
2. $a = 0.3$
3. $a = 0.5$

3.6.1

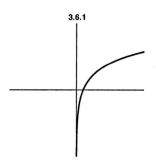

3.6.2

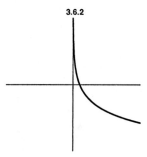

3.6.3

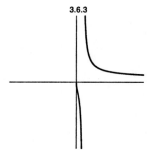

3.6.4

$\frac{1}{2}$
3

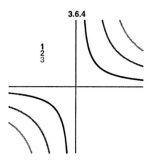

3.6.5 $y = 0.5 \ln(x^2 + a^2)$

      1. $a=0.5$
      2. $a=1.0$
      3. $a=2.0$

3.6.6 $y = 0.25 \ln[10(x^2 - a^2)]$

      1. $a=0.1$
      2. $a=0.3$
      3. $a=0.5$

3.6.7 $y = 0.5 \ln[x + (x^2 + a^2)^{1/2}]$

      1. $a=0.1$
      2. $a=0.3$
      3. $a=0.5$

3.6.8 $y = 0.5 \ln[x + (x^2 - a^2)^{1/2}]$

      1. $a=0.1$
      2. $a=0.3$
      3. $a=0.5$

3.6.5

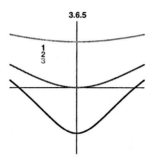

3.6.6

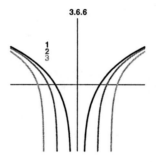

3.6.7

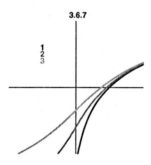

3.6.8

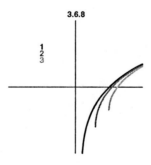

3.7 Exponential Functions

3.7.1 $y = ce^{ax}$

        1. $a=1.0, c=0.1$
        2. $a=2.0, c=0.1$
        3. $a=3.0, c=0.1$

        4. $a=-1.0, c=1.0$
        5. $a=-2.0, c=1.0$
        6. $a=-3.0, c=1.0$

3.7.2 $y = 1/(a + be^{cx})$

Special case: $a=1$, $b=1$ gives *sigmoidal curve*.

        1. $a=1.0, b=1.0, c=2.0$
        2. $a=1.0, b=1.0, c=4.0$
        3. $a=1.0, b=1.0, c=6.0$

        4. $a=-2.0, b=1.0, c=4.0$
        5. $a=-4.0, b=1.0, c=4.0$
        6. $a=-6.0, b=1.0, c=4.0$

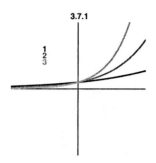

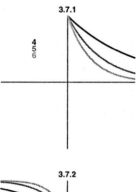

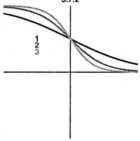

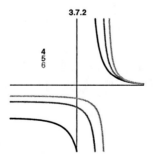

3.7.3 $y = ae^{bx} + ce^{dx}$

        1. $a=1.0, b=1.0, c=-1.0, d=3.0$
        2. $a=1.0, b=1.0, c=-1.0, d=4.0$
        3. $a=1.0, b=1.0, c=-1.0, d=5.0$

        4. $a=0.1, b=2.0, c=0.1, d=-1.0$
        5. $a=0.1, b=2.0, c=0.1, d=-2.0$
        6. $a=0.1, b=2.0, c=0.1, d=-3.0$

3.7.4 $y = 1/(ae^{bx} + ce^{dx})$

        1. $a=10.0, b=1.0, c=-10.0, d=2.0$
        2. $a=10.0, b=1.0, c=-10.0, d=3.0$
        3. $a=10.0, b=1.0, c=-10.0, d=6.0$

        4. $a=1.0, b=4.0, c=1.0, d=-1.0$
        5. $a=1.0, b=4.0, c=1.0, d=-2.0$
        6. $a=1.0, b=4.0, c=1.0, d=-4.0$

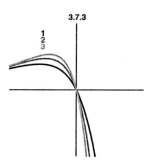

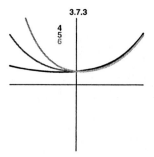

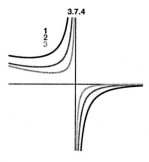

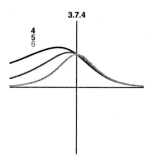

3.7.5 $y = ce^{ax^2}$

$a < 1$ gives *Gaussian curve* (also called *normal curve*)

        1. $a = -1.0, c = 1.0$
        2. $a = -2.0, c = 1.0$
        3. $a = -4.0, c = 1.0$

        4. $a = 1.0, c = 0.3$
        5. $a = 2.0, c = 0.3$
        6. $a = 4.0, c = 0.3$

3.7.6 $y = ce^{1/ax}$

        1. $a = 1.0, c = 0.2$
        2. $a = 2.0, c = 0.2$
        3. $a = 4.0, c = 0.2$

3.7.7 $y = ce^{1/(ax^2)}$

        1. $a = 1.0, c = 0.05$
        2. $a = 2.0, c = 0.05$
        3. $a = 4.0, c = 0.05$

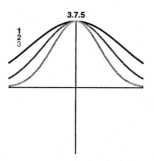

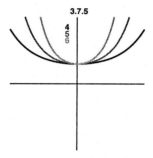

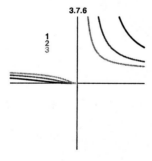

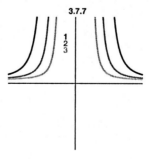

3.7.8 $y = ce^{1/(1-ax^2)}$

          1. $a=2.0, c=0.1$
          2. $a=4.0, c=0.1$
          3. $a=8.0, c=0.1$

3.7.9 $y = ce^{1/(1-a|x|)}$

          1. $a=2.0, c=0.1$
          2. $a=4.0, c=0.1$
          3. $a=8.0, c=0.1$

3.7.10 $y = c(1 + e^{ax})/(1 - e^{bx})$

          1. $a=1.0, b=1.0, c=0.02$
          2. $a=3.0, b=1.0, c=0.02$
          3. $a=4.0, b=1.0, c=0.02$

          4. $a=1.0, b=1.0, c=0.02$
          5. $a=1.0, b=3.0, c=0.02$
          6. $a=1.0, b=6.0, c=0.02$

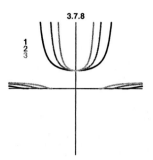

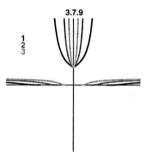

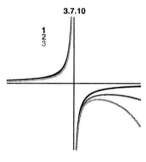

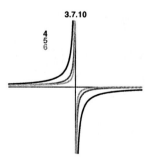

3.8 Hyperbolic Functions

3.8.1 $y = 0.1 \sinh(5x)$

3.8.2 $y = 0.1 \cosh(5x)$

Catenary

3.8.3 $y = \tanh(5x)$

3.8.4 $y = 0.1 \coth(5x)$

3.8.1

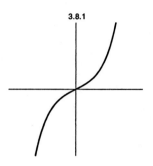

3.8.2

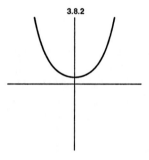

3.8.3

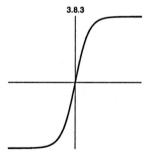

3.8.4

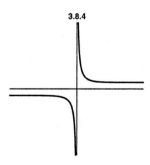

3.8.5 $y = \text{sech}(5x)$

3.8.6 $y = 0.1\,\text{csch}(5x)$

3.8.7 $y = \sinh^2(x)$

3.8.8 $y = 0.5\cosh^2(x)$

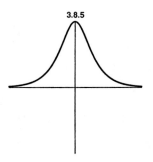

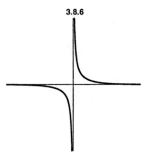

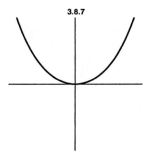

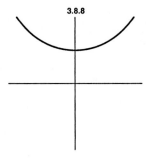

3.8.9 $y = \tanh^2(5x)$

3.8.10 $y = 0.25/[\sinh(x)\cosh(x)]$

3.8.11 $y = \sinh(ax)\cosh(bx)$

                  1. $a=1.0, b=0.5$
                  2. $a=1.0, b=2.0$
                  3. $a=1.0, b=4.0$

3.8.12 $y = \sinh(ax)\sinh(bx)$

                  1. $a=1.0, b=1.5$
                  2. $a=1.0, b=2.0$
                  3. $a=1.0, b=3.0$

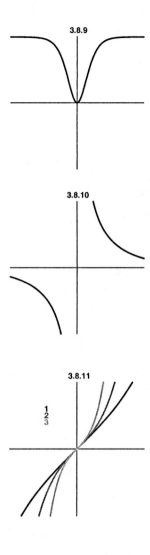

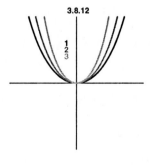

3.8.13 $y = 0.5 \cosh(ax) \cosh(bx)$

          1. $a=1.0, b=1.2$
          2. $a=1.0, b=2.0$
          3. $a=1.0, b=4.0$

3.9 Inverse Hyperbolic Functions

3.9.1 $y = 0.5 \sinh^{-1}(5x)$

3.9.2 $y = 0.5 \cosh^{-1}(5x)$

3.9.3 $y = 0.2 \tanh^{-1}(x)$

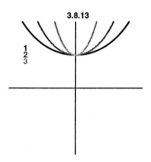

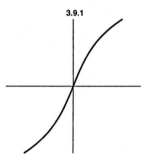

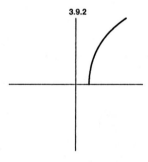

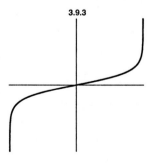

3.9.4 $y = \coth^{-1}(5x)$

3.9.5 $y = 0.2\,\text{sech}^{-1}(x)$

3.9.6 $y = 0.2\,\text{csch}^{-1}(x)$

3.10 Trigonometric and Exponential Functions Combined

3.10.1 $y = e^{ax}\sin(2\pi bx)$

          1. $a = -1.0, b = 4.0$
          2. $a = -2.0, b = 4.0$

3.9.4

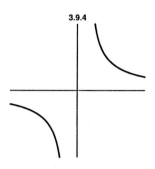

3.9.5

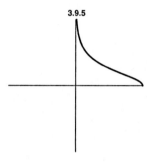

3.9.6

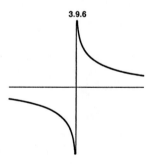

3.10.1

$\frac{1}{2}$

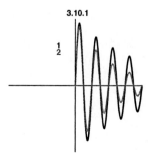

3.10.2 $y = e^{ax}\cos(2\pi bx)$

           1. $a = -1.0, b = 4.0$
           2. $a = -2.0, b = 4.0$

3.10.3 $y = 0.5\, e^{ax}/\sin(2\pi bx)$

           1. $a = -1.0, b = 4.0$
           2. $a = -2.0, b = 4.0$

3.10.4 $y = 0.5\, e^{ax}/\cos(2\pi bx)$

           1. $a = -1.0, b = 4.0$
           2. $a = -2.0, b = 4.0$

3.11 Trigonometric Functions Combined with Powers of x

3.11.1 $y = x\sin(2\pi ax)$

$a = 4.0$

3.10.2

$\frac{1}{2}$

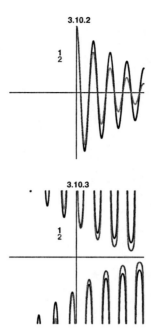

3.10.3

$\frac{1}{2}$

3.10.4

$\frac{1}{2}$

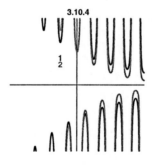

3.11.1

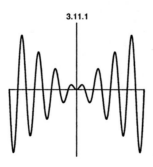

3.11.2 $y = x \cos(2\pi a x)$

$a = 4.0$

3.11.3 $y = x/\sin(2\pi a x)$

$a = 4.0$

3.11.4 $y = x/\cos(2\pi a x)$

$a = 4.0$

3.11.5 $y = \sin(2\pi a x)/(2\pi a x)$

Sinc function
$a = 4.0$

3.11.2

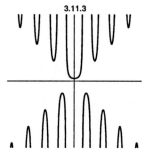

3.11.3

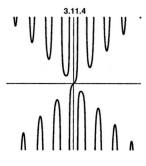

3.11.4

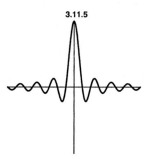

3.11.5

3.11.6 $y = \cos(2\pi ax)/(2\pi ax)$

$a = 4.0$

3.11.7 $y = x \sin^2(2\pi ax)$

$a = 4.0$

3.11.8 $y = x \cos^2(2\pi ax)$

$a = 4.0$

3.11.9 $y = 0.025 \sin(2\pi ax)/x^2$

$a = 4.0$

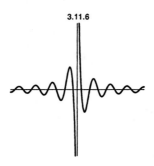

3.11.6

3.11.7

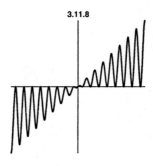

3.11.8

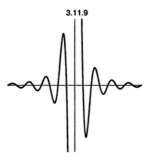

3.11.9

3.11.10 $y = 0.01 \cos(2\pi ax)/x^2$

$a = 4.0$

3.11.11 $y = 0.5\ x/\sin^2(2\pi ax)$

$a = 4.0$

3.11.12 $y = 0.5x/\cos^2(2\pi ax)$

$a = 4.0$

3.11.13 $y = 0.5x/[1 + \sin(2\pi ax)]$

$a = 4.0$

3.11.10

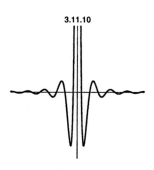

3.11.11

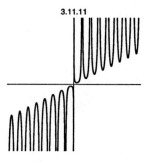

3.11.12

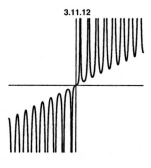

3.11.13

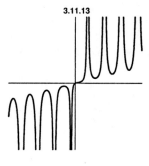

3.11.14 $y = 0.5x/[1 + \cos(2\pi ax)]$

$a = 4.0$

3.11.15 $y = 0.5x/[1 - \sin(2\pi ax)]$

$a = 4.0$

3.11.16 $y = 0.5x/[1 - \cos(2\pi ax)]$

$a = 4.0$

3.12 Logarithmic Functions Combined with Powers of x

3.12.1 $y = x \ln(ax)$

        1. $a = 1.0$
        2. $a = 2.0$
        3. $a = 4.0$

3.11.14

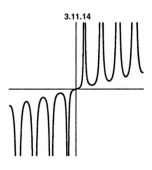

3.11.15

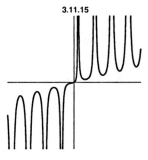

3.11.16

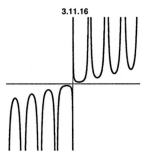

3.12.1

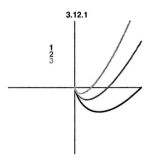

3.12.2 $y = x^2 \ln(ax)$

       1. $a=1.0$
       2. $a=2.0$
       3. $a=4.0$

3.12.3 $y = 0.05/[x \ln(ax)]$

       1. $a=1.5$
       2. $a=2.0$
       3. $a=4.0$

3.12.4 $y = 0.005/[x^2 \ln(ax)]$

       1. $a=1.5$
       2. $a=2.0$
       3. $a=4.0$

3.12.5 $y = 0.1 \ln(ax)/x$

       1. $a=1.0$
       2. $a=3.0$
       3. $a=9.0$

3.12.2

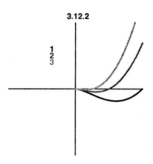

3.12.3

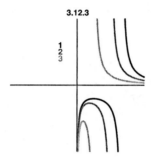

3.12.4

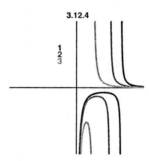

3.12.5

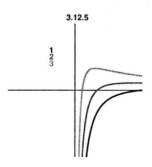

3.12.6 $y = 0.5x/\ln(ax)$

        1. $a=2.0$
        2. $a=3.0$
        3. $a=9.0$

3.12.7 $y = x \ln(ax + b)$

        1. $a=1.0, b=2.0$
        2. $a=2.0, b=2.0$
        3. $a=4.0, b=2.0$

        4. $a=3.0, b=-1.0$
        5. $a=4.0, b=-1.0$
        6. $a=6.0, b=-1.0$

3.12.8 $y = 0.1 \ln(ax + b)/x$

        1. $a=2.0, b=2.0$
        2. $a=3.0, b=2.0$
        3. $a=4.0, b=2.0$

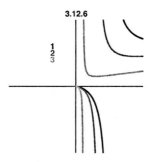

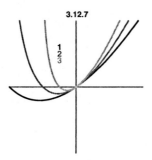

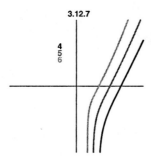

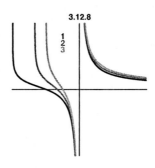

4. $a=3.0, b=-1.0$
5. $a=5.0, b=-1.0$
6. $a=9.0, b=-1.0$

3.12.9 $y = x \ln(x^2 + a^2)$

1. $a=0.0$
2. $a=0.5$
3. $a=1.0$

3.12.10 $y = 0.5x \ln(x^2 - a^2)$

1. $a=0.1$
2. $a=0.3$
3. $a=0.5$

3.13 Exponential Functions Combined with Powers of x

3.13.1 $y = 0.5\, xe^{ax}$

1. $a=1.0$
2. $a=2.0$
3. $a=3.0$

3.12.8

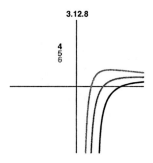

4
5
6

3.12.9

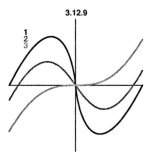

1
2
3

3.12.10

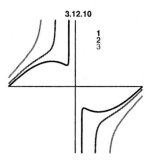

1
2
3

3.13.1

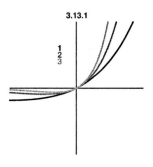

1
2
3

3.13.2 $y = x^2 e^{ax}$

      1. $a = 1.0$
      2. $a = 2.0$
      3. $a = 3.0$

3.13.3 $y = 4.0\, x^3 e^{ax}$

      1. $a = 1.0$
      2. $a = 2.0$
      3. $a = 5.0$

3.13.4 $y = 0.1\, e^{ax}/x$

      1. $a = 1.0$
      2. $a = 2.0$
      3. $a = 3.0$

3.13.5 $y = 0.03\, e^{ax}/x^2$

      1. $a = 2.0$
      2. $a = 3.0$
      3. $a = 4.0$

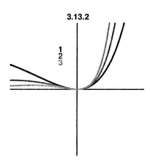

3.13.2

1
2
3

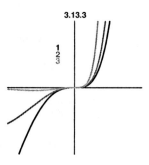

3.13.3

1
2
3

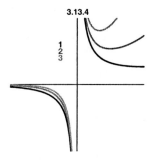

3.13.4

1
2
3

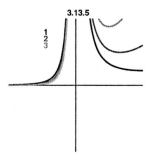

3.13.5

1
2
3

3.13.6 $y = 0.01\, e^{ax}/x^3$

       1. $a=3.0$
       2. $a=4.0$
       3. $a=5.0$

3.13.7 $y = cxe^{ax^2}$

       1. $a=-1.0,\ c=1.0$
       2. $a=-2.0,\ c=1.0$
       3. $a=-3.0,\ c=1.0$

       4. $a=1.0,\ c=0.1$
       5. $a=2.0,\ c=0.1$
       6. $a=3.0,\ c=0.1$

3.13.8 $y = cx^2 e^{ax^2}$

       1. $a=-1.0,\ c=2.0$
       2. $a=-2.0,\ c=2.0$
       3. $a=-3.0,\ c=2.0$

3.13.6

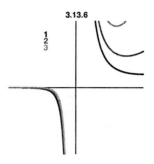

3.13.7

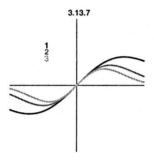

3.13.7

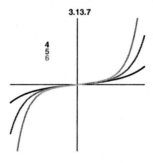

3.13.8

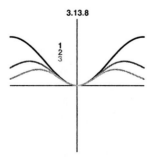

4. $a=1.0, c=0.5$
5. $a=2.0, c=0.5$
6. $a=3.0, c=0.5$

3.14   Hyperbolic Functions Combined with Powers of x

3.14.1 $y = 0.1x \sinh(5x)$

3.14.2 $y = 0.1x \cosh(5x)$

3.14.3 $y = x \tanh(5x)$

3.13.8

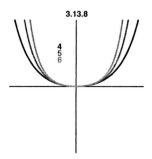

3.14.1

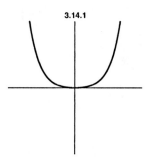

3.14.2

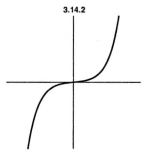

3.14.3

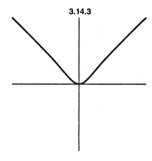

3.14.4 $y = 0.02\sinh(5x)/x$

3.14.5 $y = 0.02\cosh(5x)/x$

3.14.6 $y = 0.2\tanh(5x)/x$

3.15 Combinations of Trigonometric Functions, Exponential Functions, and Powers of x

3.15.1 $y = 0.15xe^{ax}\sin(2\pi bx)$

        1. $a=1.0, b=4.0$
        2. $a=2.0, b=4.0$

3.14.4

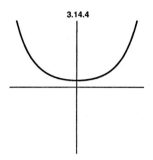

3.14.5

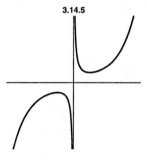

3.14.6

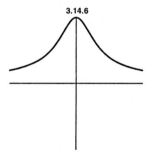

3.15.1

$\frac{1}{2}$

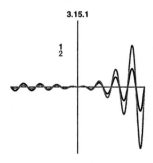

3.15.2 $y = 0.15xe^{ax}\cos(2\pi bx)$

        1. $a=1.0$, $b=4.0$
        2. $a=2.0$, $b=4.0$

3.15.3 $y = 0.1\ e^{ax}\sin(2\pi bx)/x$

        1. $a=1.0$, $b=4.0$
        2. $a=2.0$, $b=4.0$

3.15.4 $y = 0.1\ e^{ax}\cos(2\pi bx)/x$

        1. $a=1.0$, $b=4.0$
        2. $a=2.0$, $b=4.0$

3.16 Miscellaneous Functions

3.16.1 $y = a\cosh^{-1}(a/x) - (a^2 - x^2)^{1/2}$

Tractrix.

        1. $a=0.50$
        2. $a=0.75$
        3. $a=1.00$

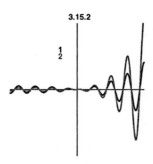

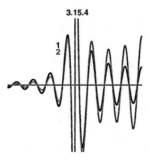

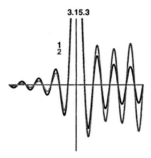

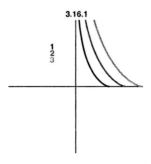

3.16.2 $y = x \cot(\pi x/2a)$

Quadratrix of Hippias

          1. $a = 0.25$
          2. $a = 0.30$
          3. $a = 0.35$

3.16.3 $y = 1 - e^{ax}$

Exponential ramp

          1. $a = -2.0$
          2. $a = -4.0$
          3. $a = -6.0$

3.16.4 $y = c(1 + 2ax^2)e^{ax^2}$

          1. $a = -3.0,\ c = 1.0$
          2. $a = -6.0,\ c = 1.0$
          3. $a = -9.0,\ c = 1.0$

          4. $a = 3.0,\ c = 0.1$
          5. $a = 6.0,\ c = 0.1$
          6. $a = 9.0,\ c = 0.1$

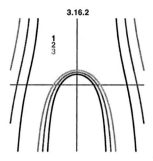

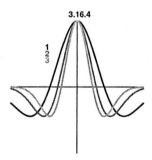

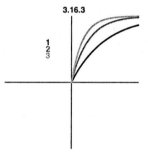

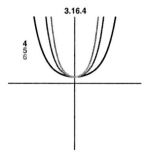

3.16.5 $\qquad y = c\arctan(e^{ax}) - b$

Special case: $a=1$, $b=\pi/2$, $c=2$ gives *Gudermannian function*

        1. $a=1.0$, $b=\pi/4$, $c=1.0$
        2. $a=3.0$, $b=\pi/4$, $c=1.0$
        3. $a=9.0$, $b=\pi/4$, $c=1.0$

3.16.6 $\qquad y = c\sin\left\{\dfrac{\pi d}{b-a}\left[\left((b-a)\dfrac{x}{d}+a\right)^{2}-a^{2}\right]\right\}$

Sweep signal (linear)
$a=5.0$, $b=25.0$, $c=0.5$, $d=1.0$

3.16.7 $\qquad y = \sin(a\pi x)\arcsin(x)$

        1. $a=1.0$
        2. $a=2.0$
        3. $a=3.0$

3.16.8 $\qquad y = c(bx)^{a\,\ln(bx)}$

        1. $a=1.0$, $b=4.0$, $c=0.1$
        2. $a=2.0$, $b=4.0$, $c=0.1$
        3. $a=4.0$, $b=4.0$, $c=0.1$

3.16.5

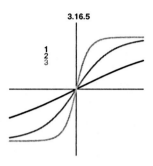

3.16.6

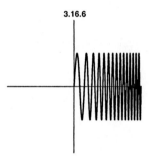

3.16.7

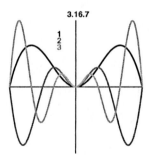

3.16.8

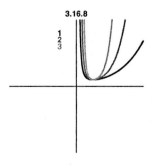

3.16.9 $y = c/(bx)^{a\,\ln(bx)}$

         1. $a=1.0, b=4.0, c=1.0$
         2. $a=2.0, b=4.0, c=1.0$
         3. $a=4.0, b=4.0, c=1.0$

3.16.10 $y = |\sin(a\pi x)|^{|\tan(a\pi x)|+1}$

$a=2.0$

3.16.11 $y = \exp\{b[\cos(a\pi x)-1]\}$

         1. $a=1.0, b=2.0$
         2. $a=1.0, b=10.0$
         3. $a=1.0, b=50.0$

         4. $a=1.0, b=4.0$
         5. $a=2.0, b=4.0$
         6. $a=4.0, b=4.0$

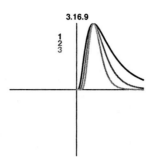

3.16.9

1
2
3

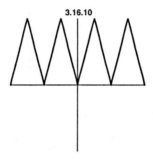

3.16.10

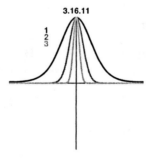

3.16.11

1
2
3

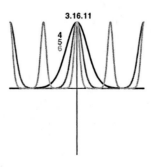

3.16.11

4
5
6

3.16.12 $y = \text{signum}[\sin(a\pi x)]|\sin(a\pi x)|^{1/b}$

$a=2.0, b=10.0$

3.16.13 $y = \sin\left(\dfrac{\pi}{2}\sin\left(\dfrac{\pi}{2}\sin\left(\cdots\dfrac{\pi}{2}\sin\left(\dfrac{a\pi x}{2}\right)\cdots\right)\right)\right)$

Let n be the order of nesting of the sine function. This function approaches a square wave as n becomes large.

$a=8.0, n=4$

3.17 Functions Expressible in Polar Coordinates

3.17.1 $r = ce^{a\theta}$ $\ln[(x^2 + y^2)/c^2] - 2a\arctan(y/x) = 0$

Logarithmic spiral (also called *equiangular spiral* or *logistique*)

a. $a=0.1, c=0.10; 0<\theta<7\pi$

b. $a=0.2, c=0.01; 0<\theta<7\pi$

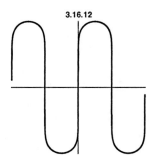

3.16.12

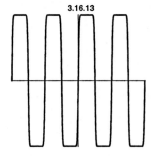

3.16.13

3.17.1a

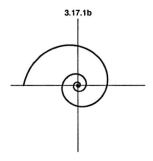

3.17.1b

3.17.2 $r = ce^{a\theta^2}$ $\ln[(x^2 + y^2)/c^2] - 2a[\arctan(y/x)]^2 = 0$

        a. $a = 0.02$, $c = 0.04$; $0 < \theta < 4\pi$

        b. $a = -0.02$, $c = 1.00$; $0 < \theta < 4\pi$

3.17.3 $r = c\cos(m\theta)$ $(x^2 + y^2)^{1/2} - c\cos[m\arctan(y/x)] = 0$

Rhodonea (also called *rose*)

        a. $c = 1.0$, $m = 4$; $0 < \theta < 2\pi$

        b. $c = 1.0$, $m = 3$; $0 < \theta < \pi$

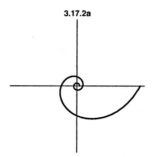

3.17.2a

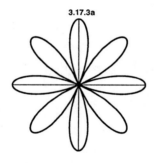

3.17.2b

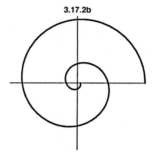

3.17.3a

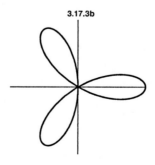

3.17.3b

3.17.4 $r = c/\cos(m\theta)$ $(x^2 + y^2)^{1/2} - c/\cos[m \arctan(y/x)] = 0$

Epi-spiral

a. $c=0.1$, $m=4$; $0<\theta<2\pi$

b. $c=0.1$, $m=3$; $0<\theta<2\pi$

3.17.5 $r = c \sin(a\theta)/\theta$ $(x^2 + y^2)^{1/2}\arctan(y/x) - c \sin[a \arctan(y/x)] = 0$

Special case: $a=1$ gives *cochleoid*

a. $a=1.0$, $c=1.0$; $-6\pi<\theta<6\pi$

b. $a=2.0$, $c=0.5$; $-3\pi<\theta<3\pi$

3.17.4a

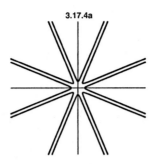

3.17.4b

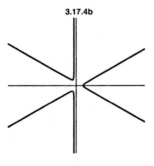

3.17.5a

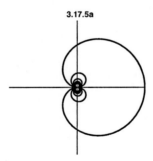

3.17.5b

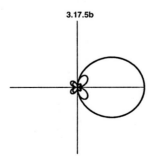

3.17.6 $r = c/\sinh(a\theta)$ $(x^2 + y^2)^{1/2} - c/\sinh[a\ \arctan(y/x)] = 0$

Spiral of Poinsot

        a. $a = 1.0;\ c = 0.5;\ -2\pi < \theta < 2\pi$

        b. $a = 0.5;\ c = 0.5,\ -4\pi < \theta < 4\pi$

3.17.7 $r = c/\cosh(a\theta)$ $(x^2 + y^2)^{1/2} - c/\cosh[a\ \arctan(y/x)] = 0$

Spiral of Poinsot

        a. $a = 1.0,\ c = 1.0;\ -2\pi < \theta < 2\pi$

        b. $a = 0.5,\ c = 1.0;\ -4\pi < \theta < 4\pi$

3.17.6a

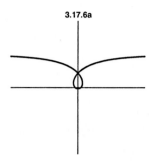

3.17.6b

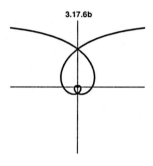

3.17.7a

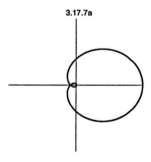

3.17.7b

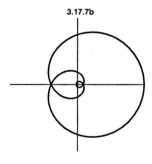

3.17.8 $r = c(2a \cos \theta + 1)$ $(x^2 + y^2 - 2acx)^2 - c^2(x^2 + y^2) = 0$

Limacon of Pascal
Special cases:

    $a = \frac{1}{2}$ gives *cardioid*
    $a = 1$ gives *trisectrix*

        1. $a = 0.25$, $c = 0.3$; $0 < \theta < 2\pi$
        2. $a = 0.50$, $c = 0.3$; $0 < \theta < 2\pi$
        3. $a = 1.00$, $c = 0.3$; $0 < \theta < 2\pi$

3.17.9 $r^2 = c^2 \cos(2\theta)$ $(x^2 + y^2)^2 - c^2|x^2 - y^2| = 0$

Lemniscate of Bernoulli
$c = 1.0$; $-\pi/4 < \theta < \pi/4$

3.17.10 $r = c \cot \theta$ $(x^2 + y^2)y^2 - a^2x^2 = 0$

Kappa curve
$c = 0.6$; $0 < \theta < \pi$

3.17.11 $r^2 = c^2(1 - a^2 \sin^2 \theta)$ $(x^2 + y^2)^2 - c^2[x^2 + (1 - a^2)y^2] = 0$

Hippopede curve

        1. $a = 0.70$, $c = 1.0$; $0 < \theta < 2\pi$
        2. $a = 0.85$, $c = 1.0$; $0 < \theta < 2\pi$
        3. $a = 1.00$, $c = 1.0$; $0 < \theta < 2\pi$

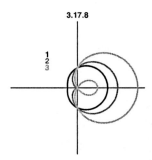

3.17.8

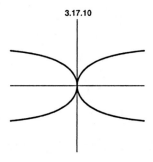

3.17.9

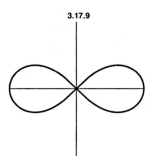

3.17.10

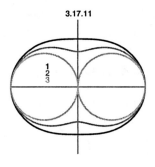

3.17.11

3.17.12 $r^2 = \dfrac{a^2\sin^2\theta - b^2\cos^2\theta}{\sin^2\theta - \cos^2\theta}$ $y^2(y^2 - a^2) - x^2(x^2 - b^2) = 0$

Devil's curve

> 1. $a = 0.25$, $b = 0.8$; $0 < \theta < 2\pi$
> 2. $a = 0.50$, $b = 0.8$; $0 < \theta < 2\pi$
> 3. $a = 0.75$, $b = 0.8$; $0 < \theta < 2\pi$

3.17.13 $r = \cos\theta(4a\sin^2\theta - b)$ $y^2[1 + (b - 4a)x] + x^2(1 + b) = 0$

Folium

> 1. $a = 0.25$, $b = 1.0$; $0 < \theta < \pi$
> 2. $a = 0.50$, $b = 1.0$; $0 < \theta < \pi$
> 3. $a = 1.00$, $b = 1.0$; $0 < \theta < \pi$

3.17.14 $r = c\sin\theta\cos^2\theta$ $(x^2 + y^2)^2 - cx^2y = 0$

Bifolia (also called *double folium*)
$c = 3.0$; $0 < \theta < \pi$

3.17.15 $r^2 = [b^4 - a^4\sin^2(2\theta)]^{1/2} + a^2\cos(2\theta)$ $[(a^2 - (x^2 + y^2)]^2 + 4a^2y^2 - b^4 = 0$

Cassinian oval
Special case: $a = b$ gives *lemniscate of Bernoulli*

> 1. $a = 0.30$, $b = 0.6$; $0 < \theta < 2\pi$
> 2. $a = 0.50$, $b = 0.6$; $0 < \theta < 2\pi$
> 3. $a = 0.60$, $b = 0.6$; $0 < \theta < 2\pi$

3.17.12

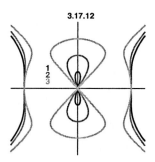

3.17.13

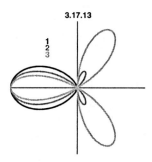

3.17.14

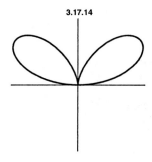

3.17.15

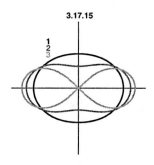

3.17.16 $\qquad r = c[1 + 2\sin(\theta/2)] \qquad (x^2 + y^2)(x^2 + y^2 + c^2 - 2c)^2 - [2c(x^2 + y^2) - 2cx] = 0$

Nephroid of Freeth
$c = 0.3;\ 0 < \theta < 4\pi$

3.17.17 $\qquad r = c\cos^3(\theta/3) \qquad 4(x^2 + y^2 - cx)^3 - 27c^2(x^2 + y^2)^2 = 0$

Cayley's sextet
$c = 1.0;\ 0 < \theta < 3\pi$

3.17.18 $\qquad r = c\dfrac{1 - a\cos\theta}{1 + a\cos\theta} \qquad (x^2 + y^2 + acx)^2 - (x^2 + y^2)(c - ax)^2 = 0$

1. $a = 0.4,\ c = 0.3;\ 0 < \theta < 2\pi$
2. $a = 0.7,\ c = 0.3;\ 0 < \theta < 2\pi$
3. $a = 1.0,\ c = 0.3;\ 0 < \theta < 2\pi$

4. $a = 2.0,\ c = 0.3;\ 0 < \theta < 2\pi$
5. $a = 4.0,\ c = 0.3;\ 0 < \theta < 2\pi$
6. $a = 9.0,\ c = 0.3;\ 0 < \theta < 2\pi$

3.17.16

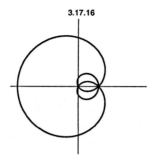

3.17.17

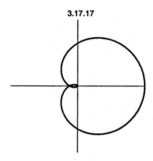

3.17.18

$\frac{1}{2}$
$\frac{3}{}$

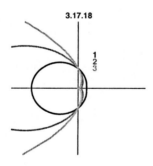

3.17.18

$\frac{4}{5}$
$\frac{6}{}$

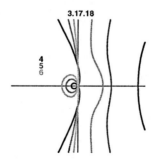

3.17.19 $r = c(1 - \tan^2\theta)$ $x^2(x^2 + y^2) + c(x^2 - y^2) = 0$

Bow curve
$c = 1.0; \ 0 < \theta < 2\pi$

3.17.20 $r = c[\cos^n\theta + 1]$

        1. $c = 0.5, n = 2; \ 0 < \theta < 2\pi$
        2. $c = 0.5, n = 10; \ 0 < \theta < 2\pi$
        3. $c = 0.5, n = 100; \ 0 < \theta < 2\pi$

        4. $c = 0.5, n = 3; \ 0 < \theta < 2\pi$
        5. $c = 0.5, n = 11; \ 0 < \theta < 2\pi$
        6. $c = 0.5, n = 101; \ 0 < \theta < 2\pi$

3.18 Functions Expressed Parametrically

3.18.1 $x = \sin(at + b\pi); y = \sin(t)$

Lissajous curves (also called *Bowditch curves*)
Let $d =$ denominator of the parameter a.

        a. $a = \frac{1}{2}, b = 0; \ 0 < t < 2\pi d$

3.17.19

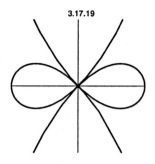

3.17.20

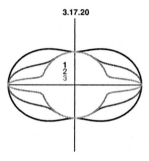

3.17.20

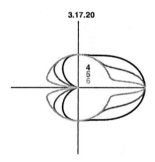

3.18.1a

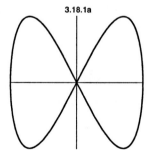

b. $a = \frac{1}{3}$, $b = 0$; $0 < t < 2\pi d$

c. $a = \frac{1}{4}$, $b = 0$; $0 < t < 2\pi d$

d. $a = \frac{2}{3}$, $b = 0$; $0 < t < 2\pi d$

e. $a = \frac{3}{4}$, $b = 0$; $0 < t < 2\pi d$

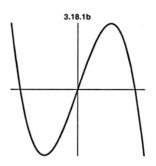

3.18.1b

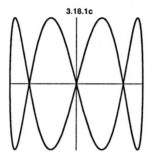

3.18.1c

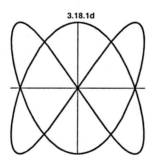

3.18.1d

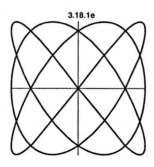

3.18.1e

f. $a=\frac{1}{2}$, $b=\frac{1}{4}$; $0<t<2\pi d$

g. $a=\frac{1}{3}$, $b=\frac{1}{4}$; $0<t<2\pi d$

3.18.2 $x = \cos(t); y = \sin(t) \sin(t/2)^n$

Teardrop curve

1. $n=1$; $0<t<2\pi$
2. $n=2$; $0<t<2\pi$
3. $n=3$; $0<t<2\pi$

3.18.3 $x = at - b\sin(t); y = c[a - b\cos(t)]$

Cycloid

a. $a=1/(4\pi)$, $b=a$, $c=2.0$; $-1/a<t<1/a$ (*ordinary cycloid* ($a=b$))

3.18.1f

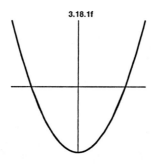

3.18.1g

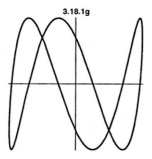

3.18.2

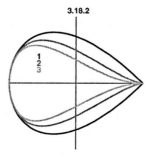

3.18.3a

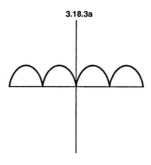

b. $a=1/(4\pi)$, $b=2a$, $c=2.0$; $-1/a<t<1/a$ *(prolate cycloid ($a<b$))*

c. $a=1/(4\pi)$, $b=a/2$, $c=2.0$; $-1/a<t<1/a$ *(curtate cycloid ($a>b$))*

3.18.4 $x = d\{(a-b)\cos(t) + c\cos[(a-b)t/b]\}; y = d\{(a-b)\sin(t) - c\sin[(a-b)t/b]\}$

Hypotrochoid

a. $a=3.0$, $b=1.0$, $c=3.0$, $d=0.15$; $0<t<2\pi$

b. $a=4.0$, $b=1.0$, $c=3.0$, $d=0.15$; $0<t<2\pi$

3.18.3b

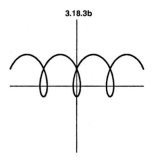

3.18.3c

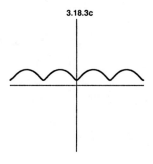

3.18.4a

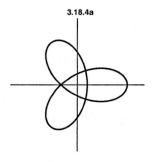

3.18.4b

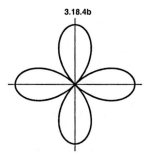

c. $a=3.0$, $b=1.0$, $c=2.0$, $d=0.25$; $0<t<2\pi$

d. $a=4.0$, $b=1.0$, $c=2.0$, $d=0.20$; $0<t<2\pi$

e. $a=3.0$, $b=1.0$, $c=1.0$, $d=0.25$; $0<t<2\pi$ (*deltoid* or *tricuspoid*)

f. $a=4.0$, $b=1.0$, $c=1.0$, $d=0.25$; $0<t<2\pi$ (*astroid*)

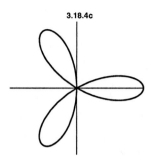

3.18.4c

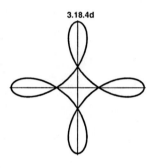

3.18.4d

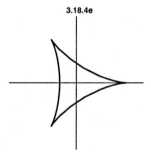

3.18.4e

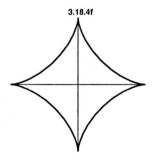

3.18.4f

3.18.5 $x = d\{(a + b)\cos(t) - c \cos[(a + b)t/b]\}; y = d\{(a + b)\sin(t) - c\sin[(a + b)t/b]\}$

Epitrochoid

   a. $a = 1.0, b = 1.0, c = 3.0, d = 0.15; 0 < t < 2\pi$

   b. $a = 2.0, b = 1.0, c = 3.0, d = 0.15; 0 < t < 2\pi$

   c. $a = 3.0, b = 1.0, c = 2.0, d = 0.15; 0 < t < 2\pi$

   d. $a = 3.0, b = 1.0, c = 5.0, d = 0.10; 0 < t < 2\pi$

3.18.5a

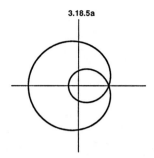

3.18.5b

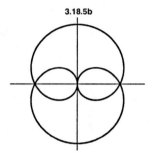

3.18.5c

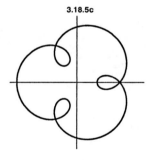

3.18.5d

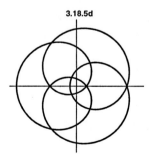

e. $a=1.0$, $b=1.0$, $c=1.0$, $d=0.25$; $0<t<2\pi$ (*cardioid*)

f. $a=2.0$, $b=1.0$, $c=1.0$, $d=0.25$; $0<t<2\pi$ (*nephroid*)

3.18.5e

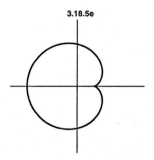

3.18.5f

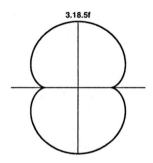

4

Polynomial Sets

The polynomial sets illustrated in this chapter are treated in detail in Abramowitz[1] or Beyer.[2] Because efficient calculation of the curves is achieved by using the recurrence relations given in these references, the relations are repeated here for anyone who may wish to generate the curves for their own purposes.

4.1 Orthogonal Polynomials

4.1.1 Legendre Polynomials $P_n(x)$

Domain: $-1<x<1$.
   Recurrence relation: $P_{n+1}(x)=[(2n+1)xP_n(x)-nP_{n-1}(x)]/(n+1)$, with $P_0(x)=1$ and $P_1(x)=x$

   0. $P_0(x)$
   1. $P_1(x)$
   2. $P_2(x)$
   3. $P_3(x)$
   4. $P_4(x)$
   5. $P_5(x)$
   6. $P_6(x)$
   7. $P_7(x)$

4.1.2 Chebyshev Polynomials of the First Kind $T_n(x)$

Domain: $-1<x<1$.
   Recurrence relation: $T_{n+1}(x)=2xT_n(x)-T_{n-1}(x)$, with $T_0(x)=1$ and $T_1(x)=x$

   0. $T_0(x)$
   1. $T_1(x)$
   2. $T_2(x)$
   3. $T_3(x)$
   4. $T_4(x)$
   5. $T_5(x)$
   6. $T_6(x)$
   7. $T_7(x)$

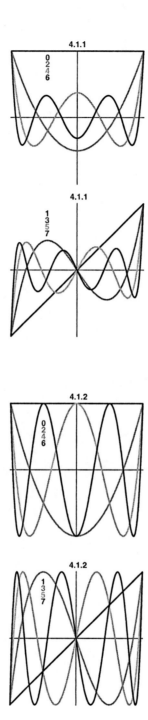

4.1.3 Chebyshev Polynomials of the Second Kind $U_n(x)$

Domain: $-1 < x < 1$.
   Recurrence relation: $U_{n+1}(x) = 2xU_n(x) - U_{n-1}(x)$, with $U_0(x) = 1$ and $U_1(x) = 2x$

0. 0.1 $U_0(x)$
1. 0.1 $U_1(x)$
2. 0.1 $U_2(x)$
3. 0.1 $U_3(x)$
4. 0.1 $U_4(x)$
5. 0.1 $U_5(x)$
6. 0.1 $U_6(x)$
7. 0.1 $U_7(x)$

4.1.4 Generalized Laguerre Polynomials $L_n^a(x)$

$a = 0$ gives ordinary Laguerre polynomials.
   Domain: $x > 0$.
   Recurrence   relation: $L_{n+1}^a(x) = [(2n + a + 1 - x)L_n^a(x) - (n + a)L_{n-1}^a(x)]/(n + 1)$,   with
$L_0^a(x) = 1$ and $L_1^a(x) = 1 - x + a$

0a. 0.1 $L_0^1(10x)$
1a. 0.1 $L_1^1(10x)$
2a. 0.1 $L_2^1(10x)$
3a. 0.1 $L_3^1(10x)$
4a. 0.1 $L_4^1(10x)$

0b. 0.1 $L_0^2(10x)$
1b. 0.1 $L_1^2(10x)$
2b. 0.1 $L_2^2(10x)$
3b. 0.1 $L_3^2(10x)$
4b. 0.1 $L_4^2(10x)$

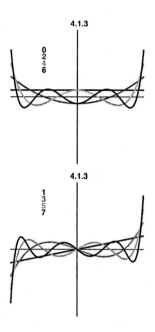

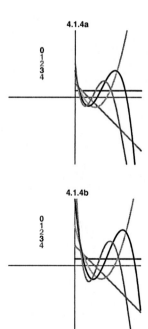

4.1.5   Laguerre Polynomials $L_n(x)$

Domain: $x > 0$.
 Recurrence relation: $L_{n+1}(x) = [(2n + 1 - x)L_n(x) - nL_{n-1}(x)]/(n + 1)$, with $L_0(x) = 1$ and $L_1(x) = 1 - x$

  0. $0.05\,L_0(10x)$
  1. $0.05\,L_1(10x)$
  2. $0.05\,L_2(10x)$
  3. $0.05\,L_3(10x)$
  4. $0.05\,L_4(10x)$
  5. $0.05\,L_5(10x)$
  6. $0.05\,L_6(10x)$
  7. $0.05\,L_7(10x)$

4.1.6   Hermite Polynomials $H_n(x)$

Domain: $x > 0$.
 Recurrence relation: $H_{n+1}(x) = 2xH_n(x) - 2nH_{n-1}(x)$, with $H_0(x) = 1$ and $H_1(x) = 2x$

  0. $0.1\,H_0(5x)$
  1. $0.1\,H_1(5x)$
  2. $0.1\,H_2(5x)/2^3$
  3. $0.1\,H_3(5x)/3^3$
  4. $0.1\,H_4(5x)/4^3$
  5. $0.1\,H_5(5x)/5^3$

4.1.7   Gegenbauer Polynomials $C_n^a(x)$

Domain: $-1 < x < 1$.
 Recurrence relation: $C_{n+1}^a(x) = [2(n + a)xC_n^a - (n + 2a - 1)C_{n-1}^a]/(n + 1)$, with $C_0^a(x) = 1$ and $C_1^a(x) = 2ax$.
 Special cases: $a = 1.0$ gives Chebyshev polynomials of the second kind; $a = 0.5$ gives Legendre polynomials

  0. $0.08\,C_0^2(x)$
  1. $0.08\,C_1^2(x)$
  2. $0.08\,C_2^2(x)$
  3. $0.08\,C_3^2(x)$
  4. $0.08\,C_4^2(x)$
  5. $0.08\,C_5^2(x)$
  6. $0.08\,C_6^2(x)$
  7. $0.08\,C_7^2(x)$

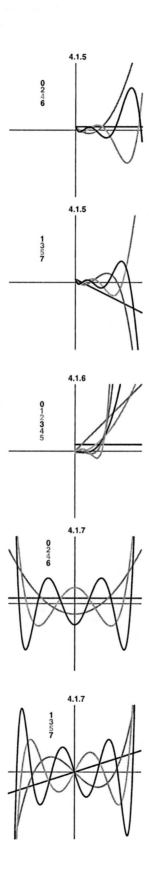

4.1.8 Jacobi Polynomials $P_n^{a,b}(x)$

Domain: $-1 < x < 1$.

Recurrence relation: $P_{n+1}^{a,b} = \{(2n + a + b + 1)[(a^2 - b^2) + (2n + a + b + 2)(2n + a + b)x]$ $P_n^{a,b} - 2(n + a)(n + b)(2n + a + b + 2)P_{n-1}^{a,b}\}/[2(n + 1)(n + a + b + 1)(2n + a + b)$, with $P_0^{a,b} = 1$ and $P_1^{a,b} = [a - b + (a + b + 2)x]/2$

0a. $P_0^{-1/2,1/2}(x)$
1a. $P_1^{-1/2,1/2}(x)$
2a. $P_2^{-1/2,1/2}(x)$
3a. $P_3^{-1/2,1/2}(x)$
4a. $P_4^{-1/2,1/2}(x)$
5a. $P_5^{-1/2,1/2}(x)$

0b. $P_0^{-1/2,1}(x)$
1b. $P_1^{-1/2,1}(x)$
2b. $P_2^{-1/2,1}(x)$
3b. $P_3^{-1/2,1}(x)$
4b. $P_4^{-1/2,1}(x)$
5b. $P_5^{-1/2,1}(x)$

0c. $P_0^{1,-1/2}(x)$
1c. $P_1^{1,-1/2}(x)$
2c. $P_2^{1,-1/2}(x)$
3c. $P_3^{1,-1/2}(x)$
4c. $P_4^{1,-1/2}(x)$
5c. $P_5^{1,-1/2}(x)$

0d. $P_0^{1,1/2}(x)$
1d. $P_1^{1,1/2}(x)$
2d. $P_2^{1,1/2}(x)$
3d. $P_3^{1,1/2}(x)$
4d. $P_4^{1,1/2}(x)$
5d. $P_5^{1,1/2}(x)$

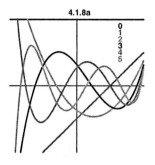

4.1.8a

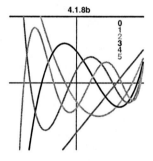

4.1.8b

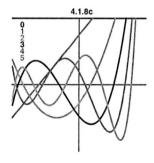

4.1.8c

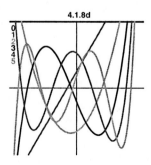

4.1.8d

4.2   Nonorthogonal Polynomials

4.2.1   Bernoulli Polynomials $B_n(x)$

Domain: $-\infty < x < \infty$.
   Recurrence relation: none.

   0. $B_0(2x)$
   1. $B_1(2x)$
   2. $B_2(2x)$
   3. $B_3(2x)$
   4. $B_4(2x)$
   5. $B_5(2x)$

4.2.2   Euler Polynomials $E_n(x)$

Domain: $-\infty < x < \infty$.
   Recurrence relation: none.

   0. $E_0(2x)$
   1. $E_1(2x)$
   2. $E_2(2x)$
   3. $E_3(2x)$
   4. $E_4(2x)$
   5. $E_5(2x)$

4.2.3   Neumann Polynomials $O_n(x)$

Domain: $x > 0$.
   Recurrence relation (for $n > 1$): $O_{n+1}(x) = (n+1)(2/x)O_n(x) - [(n+1)/(n-1)]O_{n-1}(x) + (2n/x)\sin^2(n\pi/2)$, with $O_0(x) = 1/x$, $O_1(x) = 1/x^2$, and $O_2(x) = 1/x + 4/x^3$

   0. $0.05\ O_0(5x)$
   1. $0.05\ O_1(5x)$
   2. $0.05\ O_2(5x)$
   3. $0.05\ O_3(5x)$
   4. $0.05\ O_4(5x)$
   5. $0.05\ O_5(5x)$

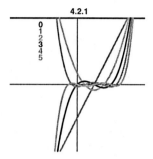

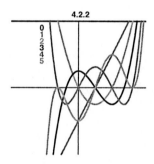

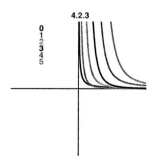

4.2.4 Schlafli Polynomials $S_n(x)$

Domain: $x > 0$.

Recurrence relation from Neumann polynomials: $S_n(x) = [2xO_n(x) - 2\cos^2(n\pi/2)]/n$, with $S_0 = 0$

1. $0.05\ S_1(5x)$
2. $0.05\ S_2(5x)$
3. $0.05\ S_3(5x)$
4. $0.05\ S_4(5x)$
5. $0.05\ S_5(5x)$

References

1. M. Abramowitz, ed. 1974. *Handbook of Mathematical Functions, with Formulas, Graphs, and Mathematical Tables*, New York: Dover.
2. W.H. Beyer, ed. 1987. *CRC Handbook of Mathematical Sciences*, 6th Ed., Boca Raton, FL: CRC Press.

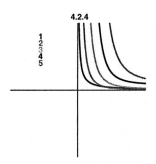

5

Special Functions in Mathematical Physics

The curves in this chapter are found in Abramowitz,[1] and the names and notation used here conform with that reference. The approximations necessary to compute these curves are also given there; for purposes of illustrating the curves, the approximations were implemented such that accuracy was attained to at least three significant figures for all plotted points of a curve. Such accuracy is sufficient for illustrative purposes and was efficiently achieved in all cases. The curves shown in this chapter are only representative; the interested reader should, when necessary, consult the above reference, or similar ones such as Jahnke and Emde,[2] Beyer,[3] and Gradshteyn and Ryzhik,[4] for a complete treatment of these curves. The reader should be aware that many of the functions are defined for a complex argument while they are usually only plotted for a real argument in this chapter, thus showing only a vertical slice of the three-dimensional surface over the complex plane.

5.1 Exponential and Related Integrals

5.1.1 Exponential Integral $E_n(x) = \int_1^\infty \frac{e^{-xt}}{t^n}\,dt$

Domain: $x > 0$.
Recurrence relation: $E_{n+1}(x) = (1/n)[e^{-x} - xE_n(x)]$ $n = 1,2,3,...$, with $E_0(x) = e^{-x}/x$

0. $E_0(x)$
1. $E_1(x)$
2. $E_2(x)$
3. $E_3(x)$
4. $E_4(x)$
5. $E_5(x)$
6. $E_6(x)$
7. $E_7(x)$

5.1.2 Exponential Integral $Ei(x) = -\int_{-x}^\infty \frac{e^{-t}}{t}\,dt$

Domain: $x > 0$.
0.5 $Ei(x)$

5.1.3 Alpha Integral $\alpha_n(x) = \int_1^\infty t^n e^{-xt}\,dt$

Domain: $x > 0$.
Recurrence relation: $\alpha_{n+1}(x) = (1/x)[e^{-x} + (n+1)\alpha_n(x)]$ $n = 0,1,2,...$, with $\alpha_0(x) = e^{-x}/x$

0. 0.2 $\alpha_0(5x)$
1. 0.2 $\alpha_1(5x)$
2. 0.2 $\alpha_2(5x)$
3. 0.2 $\alpha_3(5x)$
4. 0.2 $\alpha_4(5x)$
5. 0.2 $\alpha_5(5x)$
6. 0.2 $\alpha_6(5x)$
7. 0.2 $\alpha_7(5x)$

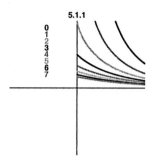

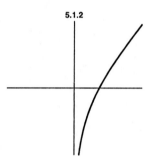

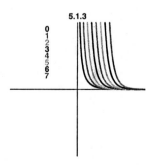

5.1.4   Beta Integral $\beta_n(x) = \int_{-1}^{1} t^n e^{-xt} dt$

Domain: $x > 0$.
Recurrence relation: $\beta_{n+1}(x) = (1/x)[(-1)^{n+1}e^x - e^{-x} + (n+1)\beta_n(x)]$ $n = 0,1,2,...,$ with
$\beta_0(x) = (2/x)\sinh(x)$

    0. $0.1\,\beta_0(5x)$
    1. $0.1\,\beta_1(5x)$
    2. $0.1\,\beta_2(5x)$
    3. $0.1\,\beta_3(5x)$
    4. $0.1\,\beta_4(5x)$
    5. $0.1\,\beta_5(5x)$
    6. $0.1\,\beta_6(5x)$
    7. $0.1\,\beta_7(5x)$

5.1.5   Logarithmic Integral $\mathrm{li}(x)$

Domain: $x > 0$.
$0.005\,\mathrm{li}(1000x)$

5.1.6   Dilogarithm $\mathrm{li}_2(x)$

Domain: $\infty < x \leq 1$.
$0.5\,\mathrm{li}_2(5x)$

5.2 Sine and Cosine Integrals

5.2.1   Sine Integral $\mathrm{Si}(x) = \int_{0}^{x} \frac{\sin t}{t} dt$

Domain: $x > 0$.
$0.5\,\mathrm{Si}(20x)$

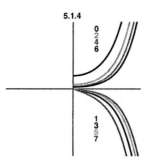

5.1.4

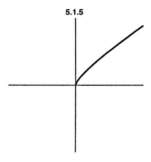

5.1.5

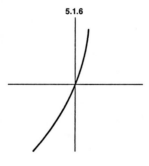

5.1.6

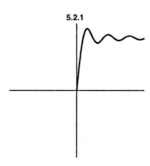

5.2.1

5.2.2   Cosine Integral $\mathrm{Ci}(x) = \gamma + \ln x + \int_0^x \dfrac{\cos t - 1}{t}\, dt$

Domain: $x > 0$.
$\mathrm{Ci}(20x)$

5.2.3   Shi Integral $\mathrm{Shi}(x) = \int_0^x \dfrac{\sinh t}{t}\, dt$

Domain: $-\infty \le x \le \infty$
$0.05\,\mathrm{Shi}(5x)$

5.2.4   Chi Integral $\mathrm{Chi}(x) = \gamma + \ln x + \int_0^x \dfrac{\cosh t - 1}{t}\, dt$

Domain: $0 < x \le \infty$.
$0.05\,\mathrm{Chi}(5x)$

5.2.5   Sici Spiral $x = c\,\mathrm{Ci}(t);$ $y = c[\mathrm{Si}(t) - \pi/2]$

$c = 0.5;\ 0.01 < t < 6\pi$

5.2.2

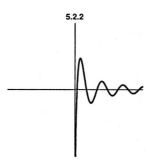

5.2.3

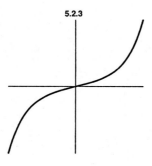

5.2.4

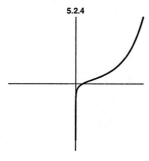

5.2.5

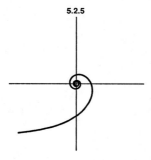

5.3   Gamma and Related Functions

5.3.1   Gamma Function $\Gamma(x) = \displaystyle\int_0^\infty t^{x-1}e^{-t}dt$

Also called *Euler's integral of the second kind*
Domain: $-\infty < x < \infty$.
Recurrence relation: $\Gamma(x+1) = x\Gamma(x)$.
$0.2\ \Gamma(5x)$

5.3.2   Complex Gamma Function $\Gamma(x + iy)$

Domain: $-\infty < x < \infty$, $\infty < y < \infty$.

   a. Contours of the surface of $|\Gamma(x+iy)|$; $-5 < x < 5$, $-1 < y < 1$

   b. Contours of the surface of $|\Gamma(x+iy)|$; $-5 < x < 5$, $-5 < y < 5$

5.3.3   Beta Function $B(x,w) = \displaystyle\int_0^1 t^{x-1}(1-t)^{w-1}dt$

Also called *Euler's integral of the first kind*.
Domain: $-\infty < x < \infty$.
Relation to Gamma Function: $B(x,w) = \Gamma(x)\Gamma(w)/\Gamma(x+w)$

   1. $0.5\ B(5x,1)$
   2. $0.5\ B(5x,2)$
   3. $0.5\ B(5x,3)$

5.3.1

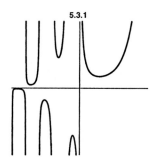

5.3.2a

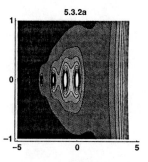

5.3.2b

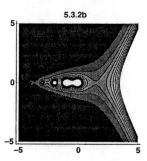

5.3.3

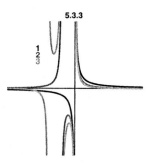

5.3.4   Psi Function $\psi(x) = [d\Gamma(x)/dx]/\Gamma(x)$

Also called *digamma function*.
Domain: $-\infty < x < \infty$.
0.2 $\psi(5x)$

5.4 Error Functions

5.4.1   Error Function $\mathrm{Erf}(x) = \dfrac{2}{\pi^{1/2}} \displaystyle\int_0^x e^{-t^2}\,dt$

Domain: $-\infty < x < \infty$.
Erf(2x)

5.4.2   Complementary Error Function $\mathrm{Erfc}(x) = 1 - Erf(x)$
Domain: $-\infty < x < \infty$.
0.5 Erfc(2x)

5.4.3   Derivatives of the Error Function $\mathrm{Erf}^{(n)}(x) = d^n[\mathrm{Erf}(x)]/dx^n$

Domain: $-\infty < x < \infty$

1. 0.40 $\mathrm{Erf}^{(1)}(2x)$
2. 0.20 $\mathrm{Erf}^{(2)}(2x)$
3. 0.05 $\mathrm{Erf}^{(3)}(2x)$

5.3.4

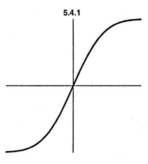

5.4.1

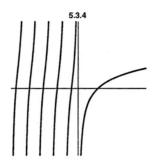

5.4.2

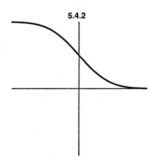

5.4.3

1
2
3

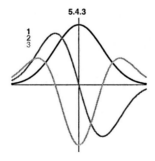

5.5   Fresnel Integrals

5.5.1   First Fresnel Integral $S(x) = \int_0^x \sin\dfrac{\pi t^2}{2}\,dt$

Domain: $-\infty < x < \infty$.
$S(5x)$

5.5.2   Second Fresnel Integral $C(x) = \int_0^x \cos\dfrac{\pi t^2}{2}\,dt$

Domain: $-\infty < x < \infty$.
$C(5x)$

5.5.3   Cornu's Spiral $x = S(t);\quad y = C(t)$

Also called *clothoid* or *Euler's spiral*.
$-6 < t < 6$

5.6   Legendre Functions

5.6.1   Associated Legendre Function of the First Kind $P_n^m(x)$

Domain: $-1 < x < 1$.
Recurrence relations: $P_{n+1}^m(x) = [(2n+1)xP_n^m - (n+m)P_{n-1}^m(x)]/(n-m+1)$ $n = 1,2,3...,$
and $P_n^{m+1}(x) = (x^2-1)^{-1/2}[(n-m)xP_n^m(x) - (n+m)P_{n-1}^m(x)]$ $m = 0,1,2,...,$ with $P_0^0 = 1$ and
$P_1^0 = x$.
Special case: $P_n^0 = $ Legendre Polynomial

   1a. $P_0^0(x)$
   2a. $P_1^0(x)$
   3a. $P_2^0(x)$
   4a. $P_3^0(x)$
   5a. $P_4^0(x)$

5.5.1

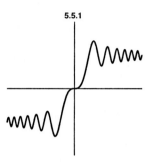

5.5.2

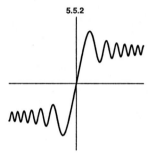

5.5.3

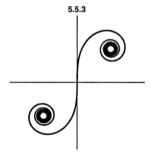

5.6.1a

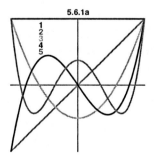

1b. $0.25P_1^1(x)$

2b. $0.25P_2^1(x)$

3b. $0.25P_3^1(x)$

4b. $0.25P_4^1(x)$

1c. $0.10P_2^2(x)$

2c. $0.10P_3^2(x)$

3c. $0.10P_4^2(x)$

1d. $0.025P_3^3(x)$

2d. $0.025P_4^3(x)$

5.6.2   Associated Legendre Function of the Second Kind $Q_n^m(x)$

Domain: $-1 < x < 1$.

Recurrence relations: $Q_{n+1}^m(x) = [(2n+1)xQ_n^m - (n+m)Q_{n-1}^m(x)]/(n-m+1)$ $n = 1,2,3...$, and $Q_n^{m+1}(x) = (x^2-1)^{-1/2}[(n-m)xQ_n^m(x) - (n+m)Q_{n-1}^m(x)]$ $m = 1,2,3...$, with $Q_0^0 = \ln[(1+x)/(1-x)]/2$ and $Q_1^0 = (x/2)\ln[(1+x)/(1-x)] - 1$

1a. $Q_0^0(x)$

2a. $Q_1^0(x)$

3a. $Q_2^0(x)$

4a. $Q_3^0(x)$

5a. $Q_4^0(x)$

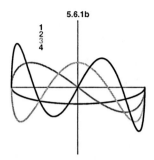

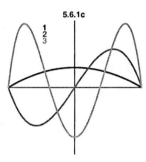

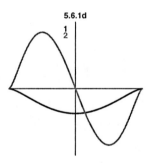

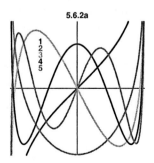

1b. $0.25Q_1^1(x)$

2b. $0.25Q_2^1(x)$

3b. $0.25Q_3^1(x)$

4b. $0.25Q_4^1(x)$

1c. $0.05Q_2^2(x)$

2c. $0.05Q_3^2(x)$

3c. $0.05Q_4^2(x)$

1d. $0.02Q_3^3(x)$

2d. $0.02Q_4^3(x)$

5.7 Bessel Functions

5.7.1   Bessel Function of the First Kind $J_n(x)$

Also called simply *Bessel function.*
Domain: $x>0$.
Recurrence relation: $J_{n+1}(x) = (2n/x)J_n(x) - J_{n-1}(x)$ $n=0,1,2,\ldots$
Symmetry: $J_{-n}(x) = (-1)^n J_n(x)$

0. $J_0(20x)$

1. $J_1(20x)$

2. $J_2(20x)$

3. $J_3(20x)$

4. $J_4(20x)$

5. $J_5(20x)$

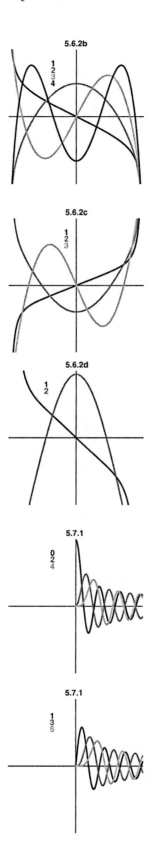

5.7.2   Bessel Function of the Second Kind $Y_n(x)$

Also called *Neumann function* or *Weber function*.
Domain: $x>0$.
Recurrence relation: $Y_{n+1}(x) = (2n/x)Y_n(x) - Y_{n-1}(x) \quad n=0,1,2\ldots$
Symmetry: $Y_{-n}(x) = (-1)^n Y_n(x)$

   0. $Y_0(20x)$
   1. $Y_1(20x)$
   2. $Y_2(20x)$
   3. $Y_3(20x)$
   4. $Y_4(20x)$
   5. $Y_5(20x)$

5.7.3   Hankel Function $H_n^{(1)}(x)$ and $H_n^{(2)}(x)$

Domain: $x>0$.
Relation to Bessel Functions: $H_n^{(1)}(x) = J_n(x) + iY_n(x)$, and $H_n^{(2)}(x) = J_n(x) - iY_n(x)$.
Recurrence relation: $H_{n+1}^{(1,2)}(x) = (2n/x)H_n^{(1,2)}(x) - H_{n-1}^{(1,2)}(x) \quad n=0,1,2\ldots$
Symmetry: $H_{-n}^{(1,2)}(x) = (-1)^n H_n^{(1,2)}(x)$

   0. $|H_0^{(m)}(20x)| \; m=1,2$
   1. $|H_1^{(m)}(20x)| \; m=1,2$
   2. $|H_2^{(m)}(20x)| \; m=1,2$
   3. $|H_3^{(m)}(20x)| \; m=1,2$
   4. $|H_4^{(m)}(20x)| \; m=1,2$
   5. $|H_5^{(m)}(20x)| \; m=1,2$

5.7.4   Complex Bessel Function $J_0(x+iy)$

Domain: $-\infty < x < \infty, \; -\infty < y < \infty$

   a. Contours of the surface $|J_0(x+iy)|$; $0 \le x \le 10, \; -2 \le y \le 2$

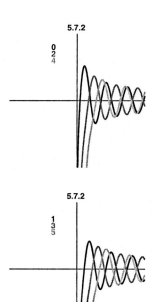

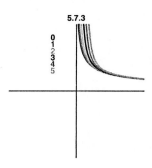

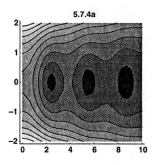

b. Contours of the surface $|J_0(x+iy)|$; $0 \leq x \leq 10$, $-5 \leq y \leq 5$

5.7.5   Bessel Function $J_n(x)$ versus Order and Argument

Domain: $-\infty < x < \infty$, $n \geq 0$.
Variable-intensity plot of $J_n(x)$: $0 \leq x \leq 25$, $0 \leq n \leq 20$.

5.8 Modified Bessel Functions

5.8.1   Modified Bessel Function $I_n(x)$

Domain: $x > 0$.
Recurrence relation: $I_{n+1}(x) = I_{n-1}(x) - (2n/x)I_n(x)$ $n = 0,1,2,\ldots$,
Symmetry: $I_{-n}(x) = I_n(x)$

   0. $0.1\, I_0(10x)$
   1. $0.1\, I_1(10x)$
   2. $0.1\, I_2(10x)$
   3. $0.1\, I_3(10x)$
   4. $0.1\, I_4(10x)$
   5. $0.1\, I_5(10x)$

5.8.2   Modified Bessel Function $K_n(x)$

Domain: $x > 0$.
Recurrence relation: $K_{n+1}(x) = K_{n-1}(x) - (2n/x)K_n(x)$ $n = 0,1,2,\ldots$,
Symmetry: $K_{-n}(x) = K_n(x)$

   0. $0.1\, K_0(5x)$
   1. $0.1\, K_1(5x)$
   2. $0.1\, K_2(5x)$
   3. $0.1\, K_3(5x)$
   4. $0.1\, K_4(5x)$
   5. $0.1\, K_5(5x)$

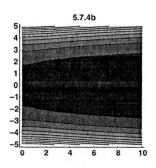

5.7.4b

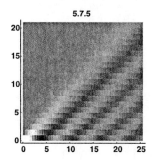

5.7.5

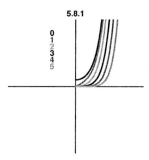

5.8.1

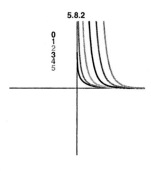

5.8.2

5.9 Kelvin Functions

5.9.1   Kelvin Function $\text{ber}_n(x)$

Domain: $x > 0$.
Recurrence relation: $\text{ber}_{n+1}(x) = -(2^{1/2}n/x)[\text{ber}_n(x) - \text{bei}_n(x)] - \text{ber}_{n-1}(x)$ $n = 1,2,3,\ldots$
Symmetry: $\text{ber}_{-n}(x) = (-1)^n\text{ber}_n(x)$

0. $0.1\,\text{ber}_0(8x)$
1. $0.1\,\text{ber}_1(8x)$
2. $0.1\,\text{ber}_2(8x)$
3. $0.1\,\text{ber}_3(8x)$
4. $0.1\,\text{ber}_4(8x)$
5. $0.1\,\text{ber}_5(8x)$

5.9.2   Kelvin Function $\text{bei}_n(x)$

Domain: $x > 0$.
Recurrence relation: $\text{bei}_{n+1}(x) = -(2^{1/2}n/x)[\text{bei}_n(x) + \text{ber}_n(x)] - \text{bei}_{n-1}(x)$ $n = 1,2,3,\ldots$
Symmetry: $\text{bei}_{-n}(x) = (-1)^n\text{bei}_n(x)$

0. $0.1\,\text{bei}_0(8x)$
1. $0.1\,\text{bei}_1(8x)$
2. $0.1\,\text{bei}_2(8x)$
3. $0.1\,\text{bei}_3(8x)$
4. $0.1\,\text{bei}_4(8x)$
5. $0.1\,\text{bei}_5(8x)$

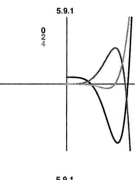

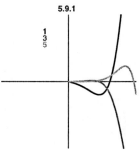

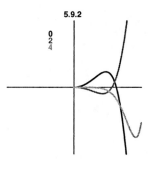

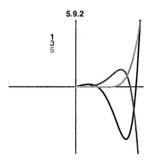

5.9.3   Kelvin Function $\ker_n(x)$

Domain: $x > 0$.
Recurrence relation: $\ker_{n+1}(x) = -(2^{1/2}n/x)[\ker_n(x) + \kei_n(x)] - \ker_{n-1}(x) \quad n = 1,2,3,\ldots$
Symmetry: $\ker_{-n}(x) = (-1)^n \ker_n(x)$

   0. $\ker_0(8x)$

   1. $\ker_1(8x)$

   2. $\ker_2(8x)$

   3. $\ker_3(8x)$

   4. $\ker_4(8x)$

   5. $\ker_5(8x)$

5.9.4   Kelvin Function $\kei_n(x)$

Domain: $x > 0$.
Recurrence relation: $\kei_{n+1}(x) = -(2^{1/2}n/x)[\kei_n(x) + \ker_n(x)] - \kei_{n-1}(x) \quad n = 1,2,3,\ldots$
Symmetry: $\kei_{-n}(x) = (-1)^n \kei_n(x)$

   0. $\kei_0(8x)$

   1. $\kei_1(8x)$

   2. $\kei_2(8x)$

   3. $\kei_3(8x)$

   4. $\kei_4(8x)$

   5. $\kei_5(8x)$

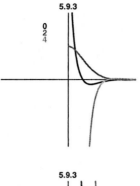

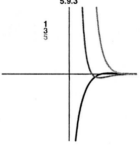

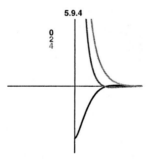

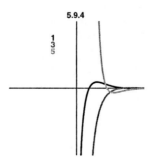

5.10   Spherical Bessel Functions

5.10.1   Spherical Bessel Function of the First Kind $j_n(x)$

Domain: $x > 0$.
Relation to Bessel Function: $j_n(x) = (\pi/2x)^{1/2} J_{n+1/2}(x)$.
Recurrence relation: $j_{n+1}(x) = [(2n+1)/x] j_n(x) - j_{n-1}(x) \quad n = 0,1,2,\ldots$
Symmetry: $j_{-n}(x) = (-1)^{-n} y_{n+1}(x)$

0. $j_0(20x)$
1. $j_1(20x)$
2. $j_2(20x)$
3. $j_3(20x)$
4. $j_4(20x)$
5. $j_5(20x)$

5.10.2   Spherical Bessel Function of the Second Kind $y_n(x)$

Domain: $x > 0$.
Relation to Bessel Function: $y_n(x) = (\pi/2x)^{1/2} Y_{n+1/2}(x)$
Recurrence relation: $y_{n+1}(x) = [(2n+1)/x] y_n(x) - y_{n-1}(x) \quad n = 0,1,2,\ldots$
Symmetry: $y_{-n}(x) = (-1)^{1-n} j_{n-1}(x)$

0. $y_0(20x)$
1. $y_1(20x)$
2. $y_2(20x)$
3. $y_3(20x)$
4. $y_4(20x)$
5. $y_5(20x)$

5.10.1

0
2
4

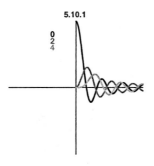

5.10.1

1
3
5

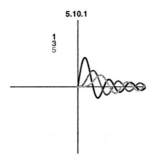

5.10.2

0
2
4

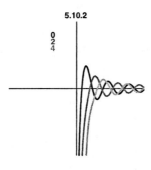

5.10.2

1
3
5

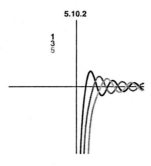

5.10.3   Spherical Bessel Function of the Third Kind $h_n^{(1)}(x)$ and $h_n^{(2)}(x)$

Domain: $x > 0$.
Relation to Hankel Function: $h_n^{(1,2)}(x) = (\pi/2x)^{1/2} H_{n+1/2}^{(1,2)}(x)$.
Recurrence relation: $h_{n+1}^{(1,2)}(x) = [(2n+1)/x]h_n^{(1,2)}(x) - h_{n-1}^{(1,2)}(x)$ $n = 0,1,2,\ldots$
Symmetry: $h_{-n}^{(1)}(x) = i(-1)^{n+1} h_{n+1}^{(2)}(x)$; $h_{-n}^{(2)}(x) = i(-1)^n h_{n+1}^{(2)}(x)$

    0. $|h_0^{(m)}(20x)|$ $m = 1,2$
    1. $|h_1^{(m)}(20x)|$ $m = 1,2$
    2. $|h_2^{(m)}(20x)|$ $m = 1,2$
    3. $|h_3^{(m)}(20x)|$ $m = 1,2$
    4. $|h_4^{(m)}(20x)|$ $m = 1,2$
    5. $|h_5^{(m)}(20x)|$ $m = 1,2$

5.11   Modified Spherical Bessel Functions

5.11.1   Modified Spherical Bessel Function of the First Kind $(\pi/2x)^{1/2} I_{n+1/2}(x)$

Domain: $x > 0$.
Recurrence relation: $I_{n+3/2}(x) = -I_{n-1/2}(x) - [(2n+1)/x)]I_{n+1/2}(x)$ $n = 0,1,2,\ldots$

    0. $0.1(\pi/20x)^{1/2} I_{1/2}(10x)$
    1. $0.1(\pi/20x)^{1/2} I_{3/2}(10x)$
    2. $0.1(\pi/20x)^{1/2} I_{5/2}(10x)$
    3. $0.1(\pi/20x)^{1/2} I_{7/2}(10x)$
    4. $0.1(\pi/20x)^{1/2} I_{9/2}(10x)$
    5. $0.1(\pi/20x)^{1/2} I_{11/2}(10x)$

5.11.2   Modified Spherical Bessel Function of the Second Kind $(\pi/2x)^{1/2} I_{-n-1/2}(x)$

Domain: $x > 0$.
Recurrence relation: $I_{-n-3/2}(x) = -I_{-n+1/2}(x) - [(2n+1)/x)]I_{-n-1/2}(x)$ $n = 0,1,2,\ldots,$

    0. $0.1(\pi/20x)^{1/2} I_{-1/2}(10x)$
    1. $0.1(\pi/20x)^{1/2} I_{-3/2}(10x)$
    2. $0.1(\pi/20x)^{1/2} I_{-5/2}(10x)$
    3. $0.1(\pi/20x)^{1/2} I_{-7/2}(10x)$
    4. $0.1(\pi/20x)^{1/2} I_{-9/2}(10x)$
    5. $0.1(\pi/20x)^{1/2} I_{-11/2}(10x)$

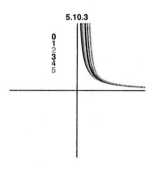

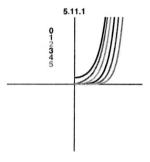

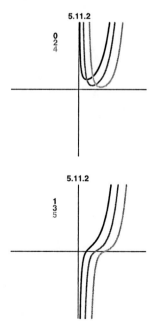

5.11.3   Modified Spherical Bessel Function of the Third Kind $(\pi/2x)^{1/2}K_{n+1/2}(x)$

Domain: $x > 0$.
Recurrence relation: $K_{n+3/2}(x) = K_{n-1/2}(x) + [(2n+1)/x)]K_{n+1/2}(x) \quad n = 0,1,2,\ldots$
Symmetry: $K_{-n-1/2}(x) = K_{n+1/2}(x)$

    0. $(\pi/20x)^{1/2}K_{1/2}(10x)$
    1. $(\pi/20x)^{1/2}K_{3/2}(10x)$
    2. $(\pi/20x)^{1/2}K_{5/2}(10x)$
    3. $(\pi/20x)^{1/2}K_{7/2}(10x)$
    4. $(\pi/20x)^{1/2}K_{9/2}(10x)$
    5. $(\pi/20x)^{1/2}K_{11/2}(10x)$

5.12   Airy Functions

5.12.1   Airy Function $Ai(x)$

Domain: $-\infty < x < \infty$
$Ai(10x)$

5.12.2   Airy Function $Bi(x)$

Domain: $-\infty < x < \infty$.
$Bi(10x)$

5.13   Riemann Functions

5.13.1   Zeta Function $\zeta(x)$

Domain: $-\infty < x < \infty$.
$0.2|\zeta(5x)|$

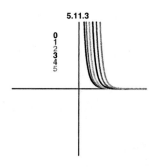

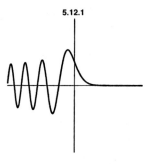

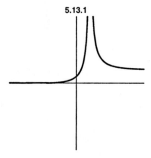

5.13.2   Zeta Function $|\zeta(1/2 + iy)|$

The line $x=1/2$ in the complex plane is the critical line of the zeta function.
Domain: $-\infty < y < \infty$ (Note: the independent variable y is along the horizontal axis.)
$0.2|\zeta(1/2 + i50y)|$

5.13.3   Complex Zeta Function $\zeta(x + iy)$
Domain: $-\infty < x < \infty$, $-\infty < y < \infty$.
Contours of the surface $|\zeta(x+iy)|$; $-4 < x < 4$, $0 < y < 15$.

5.14 Parabolic Cylinder Functions

Parabolic Cylinder Function of Half-Integer Orders Solving

5.14.1 $y'' - (x^2/4 + a)y = 0$

Domain: $x > 0$

1. $0.5\, y_0(5x)$
2. $0.5\, y_1(5x)$
3. $0.5\, y_2(5x)$
4. $0.5\, y_3(5x)$
5. $0.5\, y_4(5x)$
6. $0.5\, y_5(5x)$

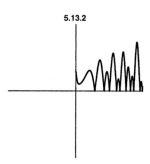

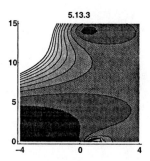

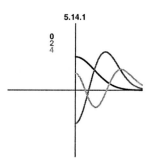

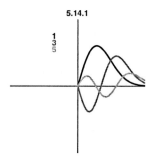

5.15   Elliptic Integrals

5.15.1   Elliptic Integral of the First Kind $F(\phi|m)$

Domain: $0 < m < 1; 0 < \phi < \pi/2$.
Contours of $F(\phi|m)$ (ϕ varies horizontally; m varies vertically).

5.15.2   Complete Elliptic Integral of the First Kind $K(m)$

Domain: $0 < m < 1$.
$0.2 \, K(m)$

5.15.3   Elliptic Integral of the Second Kind $E(\phi|m)$

Domain: $0 < m < 1; 0 < \phi < \pi/2$.
Contours of $E(\phi|m)$ (ϕ varies horizontally; m varies vertically).

5.15.4   Complete Elliptic Integral of the Second Kind $E(m)$

Domain: $0 < m < 1$.
$(2/\pi) \, E(m)$

5.15.5   Elliptic Integral of the Third Kind $\Pi(n, \phi|m)$

Domain: $0 < m < 1; 0 < \phi < \pi/2$.
Contours of $\Pi(n, \phi|m)$ for $n = 1/4$ (ϕ varies horizontally; m varies vertically).

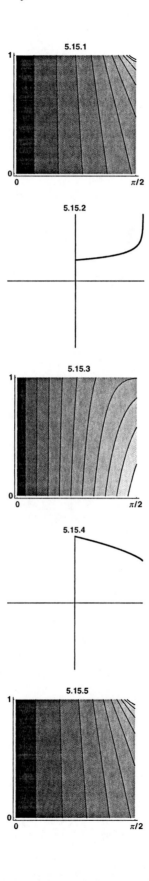

5.15.6   Complete Elliptic Integral of the Third Kind $\Pi(n, m)$

Domain: $0 < m < 1$
    1. 0.1 $\Pi(0.1, m)$
    2. 0.1 $\Pi(0.3, m)$
    3. 0.1 $\Pi(0.5, m)$
    4. 0.1 $\Pi(0.7, m)$
    5. 0.1 $\Pi(0.9, m)$

5.16 Jacobi Elliptic Functions

In the plots of the *Jacobi elliptic functions*, the independent variable u (abscissa) varies linearly from 0 to $4K(m)$, where K is the complete elliptic integral of the first kind.

5.16.1   *sn u, cn u, dn u*

    1a. *sn u*; $m = 0.25$
    2a. *cn u*; $m = 0.25$
    3a. *dn u*; $m = 0.25$

    1b. *sn u*; $m = 0.50$
    2b. *cn u*; $m = 0.50$
    3b. *dn u*; $m = 0.50$

    1c. *sn u*; $m = 0.75$
    2c. *cn u*; $m = 0.75$
    3c. *dn u*; $m = 0.75$

5.15.6

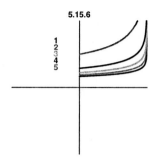

5.16.1a

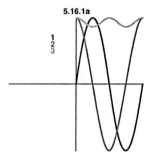

5.16.1b

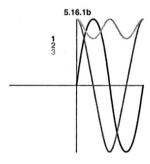

5.16.1c

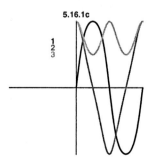

5.16.2   *sd u, cd u, nd u*

   1a. 0.5 *sd u*; $m = 0.25$
   2a. 0.5 *cd u*; $m = 0.25$
   3a. 0.5 *nd u*; $m = 0.25$

   1b. 0.5 *sd u*; $m = 0.50$
   2b. 0.5 *cd u*; $m = 0.50$
   3b. 0.5 *nd u*; $m = 0.50$

   1c. 0.5 *sd u*; $m = 0.75$
   2c. 0.5 *cd u*; $m = 0.75$
   3c. 0.5 *nd u*; $m = 0.75$

5.16.3   *sc u, dc u, nc u*

   1a. 0.5 *sc u*; $m = 0.25$
   2a. 0.5 *dc u*; $m = 0.25$
   3a. 0.5 *nc u*; $m = 0.25$

5.16.2a

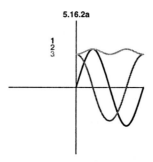

5.16.2b

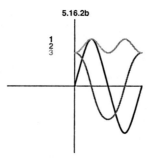

5.16.2c

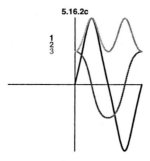

5.16.3a

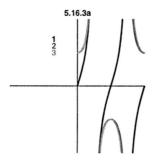

1b. 0.5 *sc u*; *m* = 0.50
2b. 0.5 *dc u*; *m* = 0.50
3b. 0.5 *nc u*; *m* = 0.50

1c. 0.5 *sc u*; *m* = 0.75
2c. 0.5 *dc u*; *m* = 0.75
3c. 0.5 *nc u*; *m* = 0.75

5.16.4   *cs u*, *ds u*, *ns u*

1a. 0.5 *cs u*; *m* = 0.25
2a. 0.5 *ds u*; *m* = 0.25
3a. 0.5 *ns u*; *m* = 0.25

1b. 0.5 *cs u*; *m* = 0.50
2b. 0.5 *ds u*; *m* = 0.50
3b. 0.5 *ns u*; *m* = 0.50

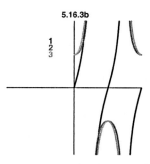

5.16.3b
1
2
3

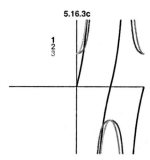

5.16.3c
1
2
3

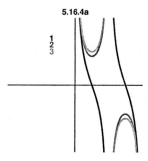

5.16.4a
1
2
3

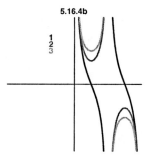

5.16.4b
1
2
3

1c. $0.5\ cs\ u;\ m=0.75$

2c. $0.5\ ds\ u;\ m=0.75$

3c. $0.5\ ns\ u;\ m=0.75$

References

1. M. Abramowitz, ed. 1974. *Handbook of Mathematical Functions, with Formulas, Graphs, and Mathematical Tables*, New York: Dover.
2. E. Jahnke and F. Emde. 1945. *Tables of Functions with Formulas and Curves*, New York, NY: Dover Publications.
3. W.H. Beyer, ed. 1987. *CRC Handbook of Mathematical Sciences*, Boca Raton, FL: CRC Press.
4. I.S. Gradshteyn, I.M. Ryzhik, A. Jeffrey, and D. Zwillinger. 2000. *Table of Integrals, Series, and Products*, 6th Ed., San Diego, USA: Academic Press.

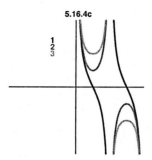

6

Green's Functions and Harmonic Functions

The *Green's function g* is the fundamental solution of a differential equation, given a Dirac delta forcing function in time and/or space. Letting D be the differential operator, the set of such equations can be written as

$$D * g = \delta(r - r_0)\delta(t - \tau)$$

where the *Dirac delta function* δ is defined to be ∞ at $r = r_0$ (one-, two-, or three-dimensional space) and $t = \tau$ (time) and zero everywhere else. Given the form of the operator D, one or the other of the delta functions may not be present. The utility of a Green's function is that after it is determined for some D, then the solution for an arbitrary forcing function can be obtained by convolution of the Green's function over the actual domain of the desired forcing function. A detailed explanation is found in standard texts.[1,2]

The Green's function may be for free space (no boundaries) or for bounded media. For 1-D, the medium is usually called a string; for 2-D, it is often called a strip or a membrane, with an implied infinitesimal thickness; a 3-D medium is often called a full-space or, if bounded on one side, a half-space. Where the medium is bounded, boundary conditions (BC) must be applied on each boundary to obtain a unique solution. The boundary conditions may be of *Dirichlet* or *Neumann* type. A Dirichlet boundary condition is where the amplitude of the boundary is fixed; a Neumann boundary condition is where the spatial derivative of the amplitude of the boundary is fixed. The form of the boundary conditions can be a scalar (usually zero) or a function along the boundary; they must, however, be continuous where boundaries meet. A third type of boundary condition, called the *mixed* or *Robin* condition, is a linear combination of the boundary amplitude and its derivative. For those Green's functions involving a time variable, the initial condition $g_{t=0} = 0$ is always applied.

Clearly, many Green's functions may be associated with a given D, depending on the nature of the boundary conditions. Here, only functions that result for unbounded media, or if bounded, from simple Dirichlet or Neumann boundary conditions, are shown. Also, because a large number of possible differential operators, D, exist, only a limited selection can be shown. The selection here concentrates on four common equations in science and engineering: (1) Poisson, (2) wave, (3) diffusion, and (4) Helmholtz. An additional equation, called the Laplacian equation ($\nabla^2 U = 0$), where δ is replaced by zero on the right-hand side, is also treated in this chapter.

Many of the curves and surfaces of this chapter are expressed as infinite sums. In plotting representative examples, a truncated series must be used. Care has been taken to use sufficient terms in the truncated series to accurately represent the function, at least outside of the immediate vicinity of the delta function that, being an inherent property of Green's functions, lies within most of the representations.

6.1 Green's Function for the Poisson Equation

The *Poisson equation*, for the Green's function g, is defined as

$$\nabla^2 g = \delta(r - r_0)$$

where ∇^2 is the Laplacian operator in one, two, or three dimensions and δ is the Dirac delta function. There is no time dependence in the Poisson equation. The source point is r_0 (in one, two, or three dimensions), and thus the delta function is non-zero only at this point. The Laplacian operator is

$$\frac{\partial^2}{\partial x^2}, \frac{\partial^2}{\partial x^2} + \frac{\partial^2}{\partial y^2}, \frac{\partial^2}{\partial x^2} + \frac{\partial^2}{\partial y^2} + \frac{\partial^2}{\partial z^2}$$

in one, two, or three dimensions. Physically, the Green's function for the Poisson equation is the static response of a medium to a permanent perturbation applied at a point.

6.1.1 1-D Bounded String

$$g(x|\xi) = \frac{2L}{\pi^2} \sum_{n=0}^{\infty} \frac{1}{n^2} \sin\left(\frac{n\pi\xi}{L}\right) \sin\left(\frac{n\pi x}{L}\right) \quad \text{Dirichlet BC}$$

$$g(x|\xi) = \frac{2L}{\pi^2} \sum_{n=0}^{\infty} \frac{1}{n^2} \cos\left(\frac{n\pi\xi}{L}\right) \cos\left(\frac{n\pi x}{L}\right) \quad \text{Neumann BC,}$$

where L is the length of the string.

a. $L = 1.0$, $\xi = 0.50$, $0 < x < L$; Dirichlet BC

b. $L = 1.0$, $\xi = 0.25$, $0 < x < L$; Dirichlet BC

6.1.1a

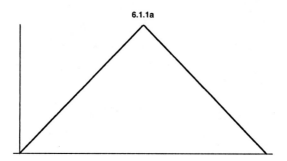

6.1.1b

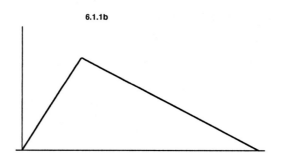

c. $L=1.0$, $\xi=0.50$, $0<x<L$; Neumann BC

d. $L=1.0$, $\xi=0.25$, $0<x<L$; Neumann BC

6.1.2 2-D Unbounded Membrane

$$g(r) = -\frac{\ln(r)}{2\pi}.$$

Domain: $0<x<4$.

6.1.3 2-D Semi-Infinite Membrane

$$g(x,y|0,\eta) = \frac{1}{2\pi}\left[\ln\left(\sqrt{x^2+(y+\eta)^2}\right) - \ln\left(\sqrt{x^2+(y-\eta)^2}\right)\right] \quad \text{Dirichlet BC}$$

$$g(x,y|0,\eta) = \frac{1}{2\pi}\left[-\ln\left(\sqrt{x^2+(y+\eta)^2}\right) - \ln\left(\sqrt{x^2+(y-\eta)^2}\right)\right] \quad \text{Neumann BC.}$$

a. $\eta=0.2$; $-1<x<1$; $0<y<1$; Dirichlet BC; viewpoint$=(10,-5,5)$

6.1.1c

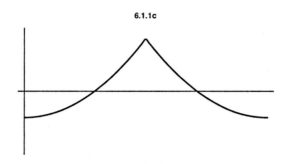

6.1.1d

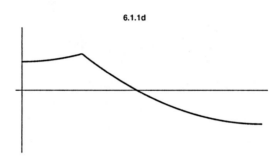

6.1.2

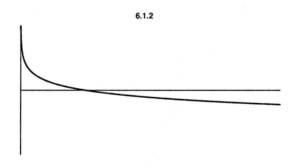

6.1.3a

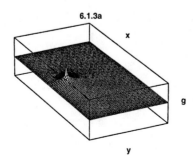

b. $\eta = 0.2$; $-1 < x < 1$; $0 < y < 1$; Neumann BC; viewpoint $= (10, -5, 5)$

6.1.4 2-D Semi-Infinite Strip

$$g(x, y | \xi, \eta) = \frac{1}{\pi} \sum_{n=1}^{\infty} \frac{1}{n} (e^{-n\pi|y-\eta|/L} - e^{-n\pi|y-\eta|/L}) \sin\left(\frac{n\pi x}{L}\right) \sin\left(\frac{n\pi \xi}{L}\right) \quad \text{Dirichlet BC}$$

$$g(x, y | \xi, \eta) = \frac{1}{\pi} \sum_{n=1}^{\infty} \frac{1}{n} (e^{-n\pi|y-\eta|/L} + e^{-n\pi|y-\eta|/L}) \sin\left(\frac{n\pi x}{L}\right) \sin\left(\frac{n\pi \xi}{L}\right) \quad \text{Neumann BC,}$$

where L is the width of the strip, truncated at $x = 0$. In both cases, Dirichlet BC are applied on the $x = 0$ and $x = L$ boundaries.

a. $L = 1.0$, $\xi = 0.5$, $\eta = 0.2$; $0 < x < L$; $0 < y < 1$; Dirichlet BC; viewpoint $= (10, -5, 5)$

b. $L = 1.0$, $\xi = 0.5$, $\eta = 0.2$; $0 < x < L$; $0 < y < 1$; Neumann BC; viewpoint $= (10, -5, 5)$

6.1.5 2-D Infinite Strip

$$g(x, y | \xi, \eta) = \sum_{n=1}^{\infty} \frac{1}{n\pi} e^{-n\pi|y|/a} \sin\left(\frac{n\pi x}{a}\right) \sin\left(\frac{n\pi \xi}{a}\right) \quad \text{Dirichlet BC}$$

$$g(x, y | \xi, \eta) = \frac{|y|}{2a} - \sum_{n=1}^{\infty} \frac{1}{n\pi} e^{-n\pi|y|/a} \cos\left(\frac{n\pi x}{a}\right) \cos\left(\frac{n\pi \xi}{a}\right) \quad \text{Neumann BC,}$$

where L is the width of the strip.

a. $L = 1.0$, $\xi = 0.0$; $0 < x < L$; $-1 < y < 1$; Dirichlet BC; viewpoint $= (10, 5, 5)$

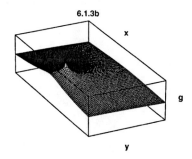

6.1.3b

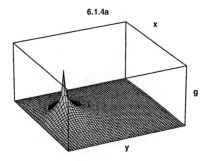

6.1.4a

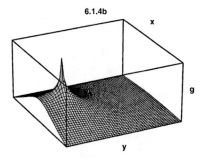

6.1.4b

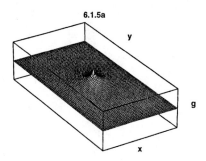

6.1.5a

b. $L=1.0$, $\xi=0.0$; $0<x<L$; $-1<y<1$; Neumann BC; viewpoint$=(10,5,5)$

6.1.6 2-D Quarter-Plane Membrane

$$g(x,y|\xi,\eta) = -\frac{1}{4\pi}\ln\left\{\frac{[(x-\xi)^2+(y-\eta)^2][(x+\xi)^2+(y+\eta)^2]}{[(x-\xi)^2+(y+\eta)^2][(x+\xi)^2+(y-\eta)^2]}\right\}\quad\text{Dirichlet BC}$$

$$g(x,y|\xi,\eta) = -\frac{1}{4\pi}\ln\{[(x-\xi)^2+(y-\eta)^2][(x+\xi)^2+(y+\eta)^2]$$

$$\times[(x+\xi)^2+(y-\eta)^2][(x-\xi)^2+(y+\eta)^2]\}\quad\text{Neumann BC.}$$

a. $\xi=0.5$, $\eta=0.5$; $0<x<1$; $0<y<1$; Dirichlet BC; viewpoint$=(10,-5,5)$

b. $\xi=0.5$, $\eta=0.5$; $0<x<1$; $0<y<1$; Neumann BC; viewpoint$=(10,-5,5)$

6.1.7 2-D Rectangular Membrane

$$g(x,y|\xi,\eta) = \frac{4}{ab}\left[\sum_{n=1}^{\infty}\sin\left(\frac{n\pi\xi}{a}\right)\sin\left(\frac{n\pi x}{a}\right)\sum_{m=1}^{\infty}\frac{\sin\left(\frac{m\pi\eta}{b}\right)\sin\left(\frac{m\pi y}{b}\right)}{(n\pi/a)^2+(m\pi/b)^2}\right]\quad\text{Dirichlet BC}$$

$$g(x,y|\xi,\eta) = \frac{4}{ab}\left[\sum_{n=1}^{\infty}\frac{\cos\left(\frac{n\pi x}{a}\right)\cos\left(\frac{n\pi\xi}{a}\right)}{2n^2\pi^2/a^2}+\sum_{m=1}^{\infty}\frac{\cos\left(\frac{m\pi y}{b}\right)\cos\left(\frac{m\pi\eta}{b}\right)}{2m^2\pi^2/b^2}\right.$$

$$\left.+\sum_{n=1}^{\infty}\cos\left(\frac{n\pi\xi}{a}\right)\cos\left(\frac{n\pi x}{a}\right)\sum_{m=1}^{\infty}\frac{\cos\left(\frac{m\pi\eta}{b}\right)\cos\left(\frac{m\pi y}{b}\right)}{(n\pi/a)^2+(m\pi/b)^2}\right]\quad\text{Neumann BC,}$$

where a is the x dimension and b is the y dimension.

a. $a=1.0$, $b=1.0$; $\xi=0.5$, $\eta=0.2$; $0<x<a$; $0<y<b$; Dirichlet BC; viewpoint$=(10,5,5)$

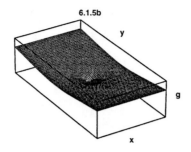

6.1.5b

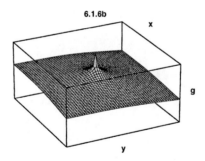

6.1.6a

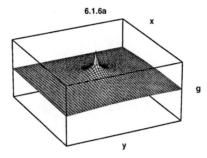

6.1.6b

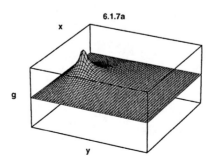

6.1.7a

b. $a=1.0$, $b=1.0$; $\xi=0.5$, $\eta=0.2$; $0<x<a$; $0<y<b$; Neumann BC; viewpoint$=$ (10,5,5)

6.1.8   2-D Circular Membrane

$$g(r,\theta|\rho,\vartheta) = \frac{1}{4\pi}\ln\left\{\frac{\rho^2[r^2+\sigma^2-2r\sigma\cos(\theta-\vartheta)]}{R^2[r^2+\rho^2-2r\rho\cos(\theta-\vartheta)]}\right\} \quad \text{Dirichlet BC}$$

$$g(r,\theta|\rho,\vartheta) = \frac{1}{4\pi}\ln\left\{\frac{[r^2+\sigma^2-2r\sigma\cos(\theta-\vartheta)][r^2+\rho^2-2r\rho\cos(\theta-\vartheta)]}{R^2r^2}\right\} \quad \text{Neumann BC,}$$

where $\sigma=R^2/\rho$ and R is the radius of the membrane.

a. $R=1.0$; $\rho=0.5$, $\vartheta=\pi/4$; $0<r<1$; $0<\theta<2\pi$; Dirichlet BC; viewpoint$=(10,-5,5)$

b. $R=1.0$; $\rho=0.5$, $\vartheta=\pi/4$; $0<r<1$; $0<\theta<2\pi$; Neumann BC; viewpoint$=(10,-5,5)$

6.1.9   3-D Unbounded Medium

$$g(r) = \frac{1}{4\pi r}.$$

Domain: $0<r<4$.

6.1.7b

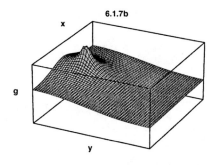

6.1.8a

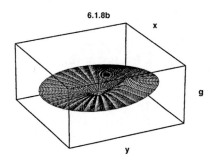

6.1.8b

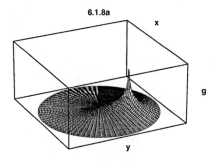

6.1.9

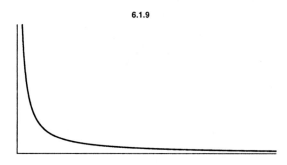

6.1.10 3-D Halfspace

$$g(r, z|0, \zeta) = \frac{1}{4\pi} \left[\frac{1}{\sqrt{r^2 + (z - \zeta)^2}} - \frac{1}{\sqrt{r^2 + (z + \zeta)^2}} \right] \quad \text{Dirichlet BC}$$

$$g(r, z|0, \zeta) = \frac{1}{4\pi} \left[\frac{1}{\sqrt{r^2 + (z - \zeta)^2}} + \frac{1}{\sqrt{r^2 + (z + \zeta)^2}} \right] \quad \text{Neumann BC.}$$

a. $\zeta = -1.0$; $-4 < r < 4$; $-4 < z < 0$; Dirichlet BC

b. $\zeta = -2.0$; $-4 < r < 4$; $-4 < z < 0$; Dirichlet BC

c. $\zeta = -1.0$; $-4 < r < 4$; $-4 < z < 0$; Neumann BC

d. $\zeta = -2.0$; $-4 < r < 4$; $-4 < z < 0$; Neumann BC

6.1.10a

6.1.10b

6.1.10c

6.1.10d

6.2   Green's Function for the Wave Equation

The *wave equation* for the Green's function, g, is defined as

$$\frac{1}{c^2}\frac{\partial^2 g}{\partial t^2} - \nabla^2 g = \delta(r - r_0)\delta(t - \tau),$$

where ∇^2 is the Laplacian operator in one, two, or three dimensions and δ is the Dirac delta function. The source point is r_0 (in one, two, or three dimensions), and thus the spatial delta function is non-zero only at this point. The source point is finite only at the time $t = \tau$ via the temporal delta function. The Laplacian operator is

$$\frac{\partial^2}{\partial x^2}, \frac{\partial^2}{\partial x^2} + \frac{\partial^2}{\partial y^2}, \frac{\partial^2}{\partial x^2} + \frac{\partial^2}{\partial y^2} + \frac{\partial^2}{\partial z^2}$$

in one, two, or three dimensions. Physically, the Green's function for the wave equation is the dynamic response of a medium to an instantaneous perturbation applied at a point.

6.2.1   1-D Unbounded String

$$g(x, t|\xi, \tau) = (c/2)H[t - \tau - |x - \xi|/c],$$

where c is the wave velocity.

$$\xi = 0, \tau = 0; -4 < x < 4, 0 < t < 4/c; \text{ viewpoint} = (-5, -7, 5)$$

6.2.2   1-D Semi-Infinite String

$$g(x, t|\xi, \tau) = (c/2)\{H[t - \tau - |x - \xi|/c] - H[t - \tau - |x + \xi|/c]\} \quad \text{Dirichlet BC}$$

$$g(x, t|\xi, \tau) = (c/2)\{H[t - \tau - |x - \xi|/c] + H[t - \tau - |x + \xi|/c]\} \quad \text{Neumann BC,}$$

where c is the wave velocity.

   a. $\xi = 1.0$, $\tau = 0$, $c = 1.0$; $0 < x < 4$, $0 < t < 4/c$; Dirichlet BC; viewpoint $= (-5, -7, 5)$

   b. $\xi = 1.0$, $\tau = 0$, $c = 1.0$; $0 < x < 4$, $0 < t < 4/c$; Neumann BC; viewpoint $= (-5, -7, 5)$

6.2.1

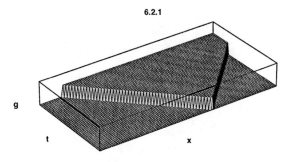

6.2.2a

6.2.2b

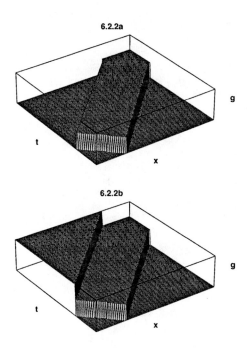

6.2.3 1-D Finite String

$$g(x,t|\xi,\tau) = \frac{2}{\pi c}H(t-\tau)\sum_{n=1}^{\infty}\frac{\sin\left(\frac{n\pi\xi}{L}\right)\sin\left(\frac{n\pi x}{L}\right)\sin\left(\frac{n\pi c(t-\tau)}{L}\right)}{n} \qquad \text{Dirichlet BC}$$

$$g(x,t|\xi,\tau) = \frac{(t-\tau)H(t-\tau)}{L} + \frac{2}{\pi c}H(t-\tau)\sum_{n=1}^{\infty}\frac{\cos\left(\frac{n\pi\xi}{L}\right)\cos\left(\frac{n\pi x}{L}\right)\sin\left(\frac{n\pi c(t-\tau)}{L}\right)}{n} \qquad \text{Neumann BC,}$$

where c is the wave velocity and L is the length of the string.

    a. $\xi=1.0$, $\tau=0$, $c=1.0$; $0<x<4$, $0<t<8/c$; Dirichlet BC; viewpoint$=(-5,-7,6)$

    b. $\xi=1.0$, $\tau=0$, $c=1.0$; $0<x<4$, $0<t<8/c$; Dirichlet BC; viewpoint$=(-5,-7,6)$

6.2.4 2-D Unbounded Membrane

$$g(r,t|0,\tau) = \frac{1}{2\pi}\frac{H\left(t-\tau-\frac{r}{c}\right)}{\sqrt{(t-\tau)^2-(r/c)^2}},$$

where c is the wave velocity.

    1. $\tau=0$, $r=0.1$, $c=1.0$; $0<t<5/c$
    2. $\tau=0$, $r=0.5$, $c=1.0$; $0<t<5/c$
    3. $\tau=0$, $r=0.9$, $c=1.0$; $0<t<5/c$

6.2.5 3-D Unbounded Medium

$$g(r,t|0,\tau) = \frac{1}{4\pi r}\delta\left(t-\tau-\frac{r}{c}\right),$$

where c is the wave velocity.

    1. $\tau=0$, $r=0.2$, $c=1.0$; $0<t<1/c$
    2. $\tau=0$, $r=0.5$, $c=1.0$; $0<t<1/c$
    3. $\tau=0$, $r=0.8$, $c=1.0$; $0<t<1/c$

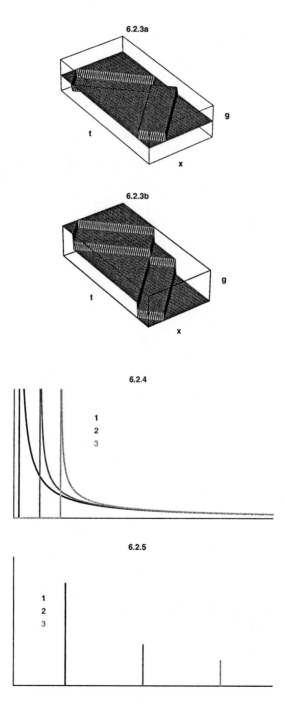

6.2.3a

6.2.3b

6.2.4

6.2.5

6.3 Green's Function for the Diffusion Equation

The *diffusion equation* (also called the *heat equation*), for the Green's function, g, is defined as

$$\frac{\partial g}{\partial t} - a^2 \nabla^2 g = \delta(r - r_0)\delta(t - \tau),$$

where ∇^2 is the Laplacian operator in one, two, or three dimensions and δ is the Dirac delta function. The source point is r_0 (in one, two, or three dimensions), and thus the delta function is non-zero only at this point. The source point is finite only at the time $t = \tau$ via the temporal delta function. The Laplacian operator is

$$\frac{\partial^2}{\partial x^2}, \frac{\partial^2}{\partial x^2} + \frac{\partial^2}{\partial y^2}, \frac{\partial^2}{\partial x^2} + \frac{\partial^2}{\partial y^2} + \frac{\partial^2}{\partial z^2}$$

in one, two, or three dimensions. Physically, the Green's function for the diffusion equation is the dynamic response of a medium to an instantaneous perturbation applied at a point.

6.3.1 1-D Unbounded String

$$g(x, t | 0, \tau) = \frac{H(t - \tau)}{\sqrt{4\pi a^2 t}} \exp\left(-\frac{x^2}{4a^2 t}\right),$$

where a^2 is the diffusivity.
$\tau = 0$, $a = 1.0$; $-1 < x < 1$, $0 < t < 1$; viewpoint $= (-5,2,3)$

6.3.2 1-D Semi-Infinite String

$$g(x, t | 0, \tau) = \frac{H(t - \tau)}{\sqrt{\pi a^2 t}} \exp\left[-\frac{(x^2 - \xi^2)}{4a^2 t}\right] \sinh\left(\frac{x\xi}{2a^2 t}\right) \quad \text{Dirichlet BC}$$

$$g(x, t | 0, \tau) = \frac{H(t - \tau)}{\sqrt{\pi a^2 t}} \exp\left[-\frac{(x^2 + \xi^2)}{4a^2 t}\right] \cosh\left(\frac{x\xi}{2a^2 t}\right) \quad \text{Neumann BC,}$$

where a^2 is the diffusivity.

a. $\tau = 0$, $\xi = 0.5$, $a = 1.0$; $0 < x < 1$, $0 < t < 1$; viewpoint $= (-5,2,3)$; Dirichlet BC

b. $\tau = 0$, $\xi = 0.5$, $a = 1.0$; $0 < x < 1$, $0 < t < 1$; viewpoint $= (-5,2,3)$; Neumann BC

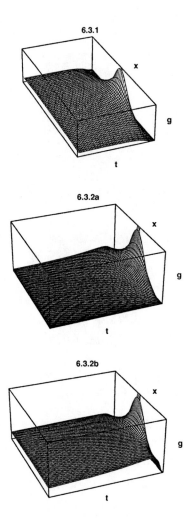

6.3.1

6.3.2a

6.3.2b

6.3.3 1-D Finite String

$$g(x,t|\xi,\tau) = \frac{2H(t-\tau)}{L} \sum_{n=1}^{\infty} \sin\left(\frac{n\pi\xi}{L}\right) \sin\left(\frac{n\pi x}{L}\right) \exp\left(-\frac{a^2 n^2 \pi^2 t}{L^2}\right) \quad \text{Dirichlet BC}$$

$$g(x,t|\xi,\tau) = \frac{2H(t-\tau)}{L} \sum_{n=1}^{\infty} \cos\left(\frac{n\pi\xi}{L}\right) \cos\left(\frac{n\pi x}{L}\right) \exp\left(-\frac{a^2 n^2 \pi^2 t}{L^2}\right) \quad \text{Neumann BC,}$$

where a^2 is the diffusivity and L is the length of the string.

    a. $\tau=0$, $\xi=0.5$, $a=1.0$, $L=1.0$; $0<x<1$, $0<t<1$; viewpoint$=(-5,2,3)$; Dirichlet BC

    b. $\tau=0$, $\xi=0.5$, $a=1.0$, $L=1.0$; $0<x<1$, $0<t<1$; viewpoint$=(-5,2,3)$; Neumann BC

6.3.4 2-D Unbounded Membrane

$$g(r,t|0,\tau) = \frac{H(t-\tau)}{4\pi a^2 t} \exp\left(-\frac{r^2}{4a^2 t}\right)$$

where a^2 is the diffusivity.

    1. $\tau=0$, $a=1.0$, $r=0.3$; $0<t<1$
    2. $\tau=0$, $a=1.0$, $r=0.6$; $0<t<1$
    3. $\tau=0$, $a=1.0$, $r=0.9$; $0<t<1$

6.3.5 2-D Circular Membrane

$$g(r,t|0,\tau) = \frac{H(t-\tau)}{\pi R^2} \sum_{n=1}^{\infty} \frac{J_0(k_{nm}r/R)}{J_1(k_{nm}^2)} \exp\frac{a^2 k_{nm}^2 t}{R^2} \quad \text{Dirichlet BC}$$

$$g(r,t|0,\tau) = \frac{H(t-\tau)}{\pi R^2} + \frac{H(t-\tau)}{\pi R^2} \sum_{n=1}^{\infty} \frac{J_0(k_{nm}r/R)}{J_0(k_{nm}^2)J_1(k_{nm}^2)} \exp\frac{a^2 k_{nm}^2 t}{R^2} \quad \text{Neumann BC,}$$

where a^2 is the diffusivity and R is the radius of the membrane. The k_{nm} are the roots of the zeroth-order Bessel function of the first kind.

    1. $\tau=0$, $a=1.0$, $R=1.0$, $r=0.3$; $0<t<1$; Dirichlet BC
    2. $\tau=0$, $a=1.0$, $R=1.0$, $r=0.6$; $0<t<1$; Dirichlet BC
    3. $\tau=0$, $a=1.0$, $R=1.0$, $r=0.9$; $0<t<1$; Dirichlet BC

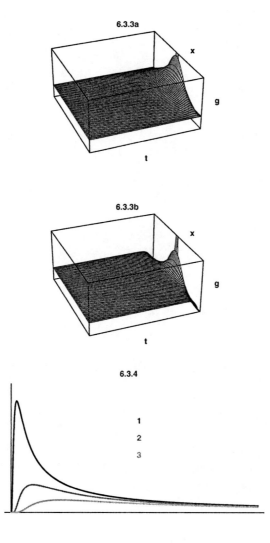

6.3.3a

6.3.3b

6.3.4

1
2
3

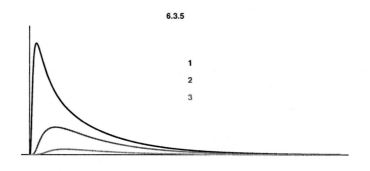

6.3.5

1
2
3

4. $\tau=0$, $a=1.0$, $R=1.0$, $r=0.3$; $0<t<1$; Neumann BC
5. $\tau=0$, $a=1.0$, $R=1.0$, $r=0.6$; $0<t<1$; Neumann BC
6. $\tau=0$, $a=1.0$, $R=1.0$, $r=0.9$; $0<t<1$; Neumann BC

6.3.6 3-D Unbounded Medium

$$g(r,t|0,\tau) = \frac{H(t-\tau)}{4\pi a^2 t^{3/2}} \exp\left(-\frac{r^2}{4a^2 t}\right),$$

where a^2 is the diffusivity.

1. $\tau=0$, $a=1.0$, $r=0.3$; $0<t<1$
2. $\tau=0$, $a=1.0$, $r=0.6$; $0<t<1$
3. $\tau=0$, $a=1.0$, $r=0.9$; $0<t<1$

6.3.7 3-D Sphere

$$g(r,t|0,\tau) = \frac{H(t-\tau)}{2rR^2} \sum_{n=1}^{\infty} n \sin\left(\frac{n\pi r}{R}\right) \exp\left(-\frac{a^2 n^2 \pi^2 t}{R^2}\right)$$

where a^2 is the diffusivity and R is the radius of the sphere.

1. $\tau=0$, $a=1.0$, $R=1.0$, $r=0.3$; $0<t<1$; Dirichlet BC
2. $\tau=0$, $a=1.0$, $R=1.0$, $r=0.6$; $0<t<1$; Dirichlet BC
3. $\tau=0$, $a=1.0$, $R=1.0$, $r=0.6$; $0<t<1$; Dirichlet BC

6.4 Green's Function for the Helmholtz Equation

The *Helmholtz equation* for the Green's function, g, is defined as

$$-\nabla^2 g - k^2 g = \delta(r - r_0),$$

where ∇^2 is the Laplacian operator in one, two, or three dimensions and δ is the Dirac delta function. There is no time dependence in the Helmholtz equation; k is the wavenumber of the spatial response, equal to frequency, in radians, divided by velocity. The source point is r_0 (in one, two, or three dimensions), and thus the delta function is non-zero only at this

6.3.5

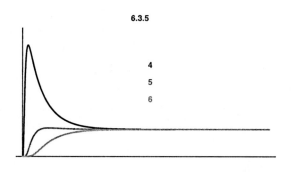

4
5
6

6.3.6

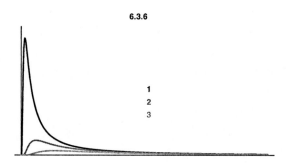

1
2
3

6.3.7

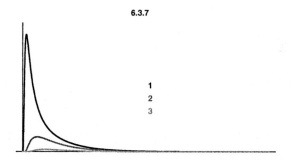

1
2
3

point. The Laplacian operator is

$$\frac{\partial^2}{\partial x^2}, \frac{\partial^2}{\partial x^2} + \frac{\partial^2}{\partial y^2}, \frac{\partial^2}{\partial x^2} + \frac{\partial^2}{\partial y^2} + \frac{\partial^2}{\partial z^2}$$

in one, two, or three dimensions. Physically, the Green's function for the Helmholtz equation is the oscillating response of a medium at a given wavenumber k to an instantaneous perturbation applied at a point.

6.4.1 1-D Unbounded String

$$g(x|0) = \frac{i}{2k} e^{ik|x|}.$$

1. $k=2\pi$; $-1<x<1$; amplitude
2. $k=4\pi$; $-1<x<1$; amplitude
3. $k=6\pi$; $-1<x<1$; amplitude
4. $k=2\pi$; $-1<x<1$; phase
5. $k=4\pi$; $-1<x<1$; phase
6. $k=6\pi$; $-1<x<1$; phase

6.4.2 1-D Finite String

$$g(x|\xi) = \frac{\sin[k, \min(x, \xi)]\sin[k(L-\max(x, \xi)]}{k\sin(kL)},$$

where k is the wavenumber (equal to frequency, in radians, divided by velocity) and L is the length of the string.

$k=1.0$, $L=10.0$; $0<x<L$, $0<\xi<L$; Dirichlet BC; viewpoint$=(-5,-7,5)$; amplitude.

6.4.3 2-D Infinite Membrane

$$g(r|0) = \frac{i}{4} H_0^{(1)}(kr),$$

where k is the wavenumber (equal to frequency, in radians, divided by velocity).

1. $k=2\pi$; $0<r<1$; amplitude
2. $k=4\pi$; $0<r<1$; amplitude
3. $k=6\pi$; $0<r<1$; amplitude

6.4.1

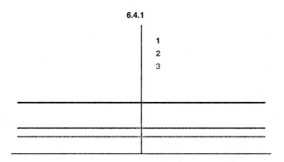

6.4.1

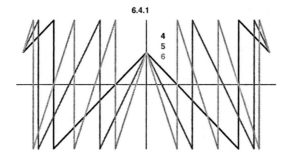

6.4.2

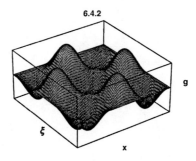

6.4.3

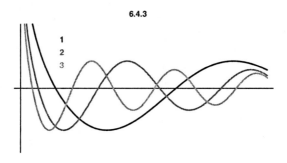

4. $k=2\pi$; $0<r<1$; phase
5. $k=4\pi$; $0<r<1$; phase
6. $k=6\pi$; $0<r<1$; phase

6.4.4 2-D Rectangular Membrane

$$g(x,y|\xi,\eta) = \frac{4}{ab} \sum_{n=1}^{\infty} \sum_{m=1}^{\infty} \frac{\sin\left(\frac{m\pi\xi}{a}\right)\sin\left(\frac{m\pi x}{a}\right)\sin\left(\frac{n\pi\eta}{b}\right)\sin\left(\frac{n\pi y}{b}\right)}{\frac{m^2\pi^2}{a^2} + \frac{n^2\pi^2}{b^2} - k^2},$$

where a is the x dimension of the rectangle and b is the y dimension of the rectangle.

a. $a=1.0$, $b=1.0$, $k=10.0$, $\xi=0.5$, $\eta=0.5$; $0<x<a$; $0<y<b$; Dirichlet BC; viewpoint$=(-5,-7,4)$; amplitude

b. $a=1.0$, $b=1.0$, $k=20.0$, $\xi=0.5$, $\eta=0.5$; $0<x<a$; $0<y<b$; Dirichlet BC; viewpoint$=(-5,-7,4)$; amplitude

c. $a=1.0$, $b=1.0$, $k=20.0$, $\xi=0.2$, $\eta=0.2$; $0<x<a$; $0<y<b$; Dirichlet BC; viewpoint$=(-5,-7,4)$; amplitude

6.4.3

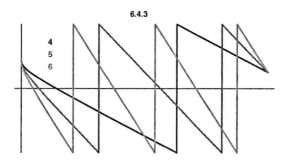

6.4.4a

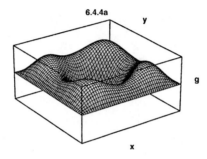

6.4.4b

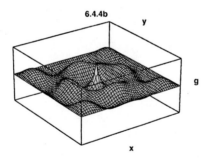

6.4.4c

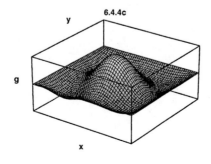

d. $a=1.0$, $b=1.0$, $k=20.0$, $\xi=0.2$, $\eta=0.2$; $0<x<a$; $0<y<b$; Dirichlet BC; viewpoint$=(-5,-7,4)$; amplitude

6.4.5   2-D Circular Membrane

$$g(r,\theta|\rho,\vartheta) = -\frac{1}{4} \sum_{n=-\infty}^{\infty} \cos[n(\theta-\vartheta)]J_n[k\min(r,\rho)]\left[Y_n[k\max(r,\rho)] - \frac{Y_n(kR)J_n[k\max(r,\rho)]}{J_n(kR)}\right]$$

where k is the wavenumber (equal to frequency, in radians, divided by velocity) and R is the radius of the membrane.

a. $R=1.0$, $k=10$; $\rho=0.0$, $\vartheta=0.0$; $0<r<1$; Dirichlet BC; viewpoint$=(10,-5,5)$; amplitude

b. $R=1.0$, $k=20$; $\rho=0.0$, $\vartheta=0.0$; $0<r<1$; Dirichlet BC; viewpoint$=(10,-5,5)$; amplitude

c. $R=1.0$, $k=10$; $\rho=0.5$, $\vartheta=0.0$; $0<r<1$; Dirichlet BC; viewpoint$=(10,-5,5)$; amplitude

d. $R=1.0$, $k=20$; $\rho=0.5$, $\vartheta=0.0$; $0<r<1$; Dirichlet BC; viewpoint$=(10,-5,5)$; amplitude

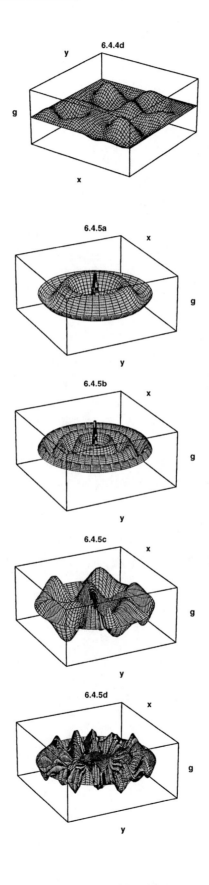

6.5 Miscellaneous Green's Functions

6.5.1 Harmonic Oscillator

The Green's function solves the *harmonic oscillator* equation:

$$a\frac{\partial^2 g}{\partial t^2} + b\frac{\partial g}{\partial t} + cg = \delta(t).$$

$$g(t|0) = \frac{e^{-\gamma t}\sin(kt)}{ak} \quad \text{underdamped}$$

$$g(t|0) = t\,e^{-\gamma t} \quad \text{critically damped}$$

$$g(t|0) = \frac{e^{-\gamma t}\sinh(kt)}{ak} \quad \text{overdamped,}$$

where the natural frequency of the system is $\omega_0 = \sqrt{c/a}$. Let $\gamma = b/(2a)$ and $k = \sqrt{\omega_0^2 - \gamma^2}$.

1. $1.0g$; $a=1.0$, $b=0.2$, $c=1.0$; $0<t<10$; underdamped
2. $1.0g$; $a=1.0$, $b=0.7$, $c=1.0$; $0<t<10$; underdamped
3. $1.0g$; $a=1.0$, $b=1.5$, $c=1.0$; $0<t<10$; underdamped
4. $2.0g$; $a=1.0$, $b=2.0$, $c=1.0$; $0<t<10$; critically damped
5. $3.0g$; $a=1.0$, $b=3.0$, $c=1.0$; $0<t<10$; overdamped
6. $3.0g$; $a=1.0$, $b=5.0$, $c=1.0$; $0<t<10$; overdamped
7. $3.0g$; $a=1.0$, $b=10.0$, $c=1.0$; $0<t<10$; overdamped

6.5.2 2-D Biharmonic Equation on Infinite Membrane

The Green's function solves the *biharmonic equation* (also called *bi-Laplacian equation*):

$$\nabla^4 g = \delta(x-\xi)\delta(y-\eta),$$

where ∇^4 is the biharmonic operator that expands to

$$\frac{\partial^4}{\partial x^4} + 2\frac{\partial^4}{\partial^2 x\partial^2 y} + \frac{\partial^4}{\partial y^4}.$$

Due to its being axisymmetric when $(\xi,\eta)=(0,0)$, the Green's function is expressed in the polar coordinate r:

$$g(r|0) = \frac{r^2}{8\pi}\ln[r-1],$$

$0<r<2.5$; viewpoint$=(-5,-7,5)$

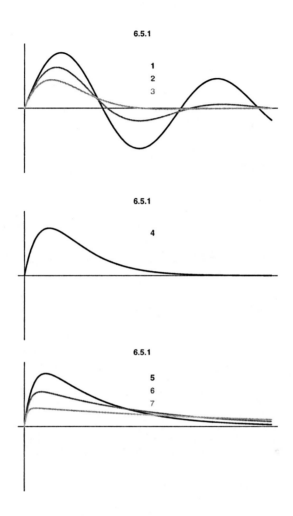

6.5.1

6.5.1

6.5.1

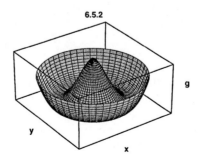

6.5.2

6.5.3   1-D Klein-Gordon Equation

The Green's function solves the 1-D *Klein-Gordon equation*:

$$\frac{1}{c^2}\left(\frac{\partial^2 g}{\partial t^2} + a^2 g\right) - \frac{\partial^2 g}{\partial x^2} = \delta(x)\delta(t)$$

$$g(x,t|0,0) = \frac{c}{2}H(ct - |x|)J_0\left(a\sqrt{t^2 - \frac{x^2}{c^2}}\right),$$

where c is the wave velocity and a^2 is the diffusivity.

$a = 2.0$, $c = 1.0$; $-1 < x < 1$, $0 < t < 1$; viewpoint $= (-5, -7, 5)$

6.6   Harmonic Functions—Solutions to Laplace's Equation

Laplace's equation is given by

$$\nabla^2 U = 0,$$

where the differential operator ∇^2 is defined as $\sum(\partial^2/\partial x_i^2)$ for $i = 1, \ldots, n$ dimensions. Functions U satisfying Laplace's differential equation are called harmonic functions.[3,4]

6.6.1   Rectangular Membrane

Let a rectangle have sides of length a in the x dimension and b in the y dimension. For Dirichlet BC on sides $x = 0$, $x = a$, $y = 0$ and for BC $= x(a - x)$ on $y = b$ (Figure a), the solution of Laplace's equation is

$$U(x,y) = \sum_{n=1}^{\infty} c_n \sinh\left(\frac{n\pi y}{a}\right)\sin\left(\frac{n\pi x}{a}\right),$$

where

$$c_n = \frac{2}{a\,\sinh(n\pi b/a)}\int_0^a x(a - x)\sin\left(\frac{n\pi x}{a}\right)\,dx.$$

For an identical BC, but now at $y = 0$, substitute $b - y$ for y. The solution for such BCs on both sides is simply the sum of the two solutions (Figure b). By suitable change of variable, the solution for BC $= y(y - b)$ on either $x = 0$ or $x = a$ can be expressed; the solution for the combination of all four non-zero BCs can then be expressed as the sum of solutions for all four separate BC's (Figure c).

  a. BC $= x(x - a)$ on $y = b$; $0 < x < 1$, $0 < y < 1$; viewpoint $= (10, 4, 4)$

  b. BC $= x(x - a)$ on $y = 0$ and $y = b$; $0 < x < 1$, $0 < y < 1$; viewpoint $= (10, 4, 4)$

  c. BC $= x(x - a)$ on $y = 0$ and $y = b$ and BC $= y(y - b)$ on $x = 0$ and $x = a$; $0 < x < 1$, $0 < y < 1$; viewpoint $= (10, 4, 4)$

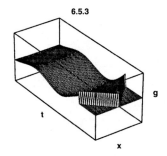

6.5.3

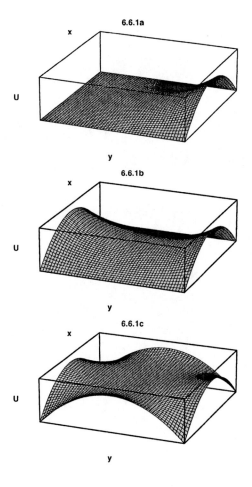

6.6.2   Circular Membrane

Assume a circular membrane of radius R. Apply the periodic BC at R of $\sin(m\theta)$. Then the solution to Laplace's equation in polar coordinates is

$$U(r, \theta) = \left(\frac{r}{R}\right)^m \sin(m\theta).$$

a. $m=1$, $R=1.0$; viewpoint$=(10, -5, 5)$

b. $m=2$, $R=1.0$; viewpoint$=(10, -5, 5)$

c. $m=4$, $R=1.0$; viewpoint$=(10, -5, 5)$

6.6.3   Annulus

Let an annulus have inside and outside radii of R_1 and R_2. Apply the BCs T_1 and T_2 on the inside and outside radii, respectively. Then the solution to Laplace's equation is the axisymmetric expression:

$$U(r) = \frac{T_2}{\ln(R_2/R_1)} \ln\left(\frac{r}{R_1}\right) + \frac{T_1}{\ln(R_1/R_2)} \ln\left(\frac{r}{R_2}\right).$$

a. $R_1=0.2$, $R_2=1.0$, $T_1=1.0$, $T_2=0.0$; viewpoint$=(10, -5, 5)$

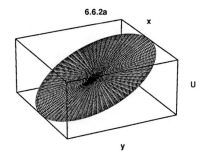

6.6.2a

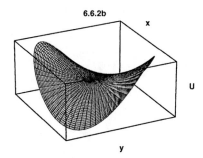

6.6.2b

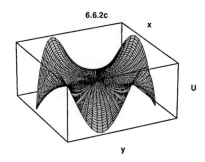

6.6.2c

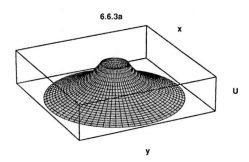

6.6.3a

b. $R_1 = 0.5$, $R_2 = 1.0$, $T_1 = 0.0$, $T_2 = 1.0$; viewpoint $= (10, -5, 5)$

References

1. D.G. Duffy. 2001. *Green's Functions with Applications*, London/Boca Raton, FL: Chapman & Hall/CRC.
2. S.I. Hayek. 2001. *Advanced Mathematical Methods in Science and Engineering*, New York: Marcel Dekker.
3. O.D. Kellogg. 1953. *Foundations of Potential Theory*, New York: Dover.
4. G.F. Carrier and C.E. Pearson. 1988. *Partial Differential Equations: Theory and Technique*, 2nd Ed., New York: Academic Press.

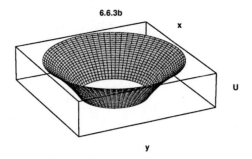

7

Special Functions in Probability and Statistics

The functions illustrated in this chapter are the common functions used in probability and statistics as presented in elementary texts on these subjects. These functions are described as either a *probability density* or a *probability distribution*. A probability density shows the weight of the possible outcomes of an experiment or set of measurements. A probability distribution is the integral of a given probability density, and it will show the accumulated probability to be unity when the entire range of possibilities is taken into account. Probability densities divide into *discrete* and *continuous* types, depending on whether the independent variable is counted in discrete units or has a continuous range of values on the real axis.

7.1 Discrete Probability Densities

The following discrete densities are plotted with the variable m on the x axis. Although a continuous line is plotted, the functions must be understood as discrete, having values only at integer m, the domain of which is listed in each case. The vertical scale is arbitrary, chosen to plot the density such that its maximum value is of nearly uniform height for all plots. Thus, the scale may change among a series of plots for a given density function. A property common to all discrete densities is that the sum over all possible m must equal unity. Therefore,

$$\sum_{m=m_1}^{m_2} P(m) = 1$$

where m_1 and m_2 are the minimum and maximum possible values of m.

7.1.1   Binomial density, $P(m|n,p) = \binom{n}{m} p^m (1-p)^{n-m}$,

where: m, number of given outcomes in n trials; n, total number of trials; p, probability of a given outcome in a single trial.

1. $n=25$, $p=0.25$; $m=0,1,2,\dots,n$
2. $n=25$, $p=0.50$; $m=0,1,2,\dots,n$
3. $n=25$, $p=0.75$; $m=0,1,2,\dots,n$

4. $n=10$, $p=0.25$; $m=0,1,2,\dots,n$
5. $n=10$, $p=0.50$; $m=0,1,2,\dots,n$
6. $n=10$, $p=0.75$; $m=0,1,2,\dots,n$

7.1.2   Geometric density, $P(m|p) = p(1-p)^{m-1}$

where: m, number of events ($m>0$); p, probability of a given event.

1. $p=0.25$; $m=1,2,3,\dots,10$
2. $p=0.50$; $m=1,2,3,\dots,10$
3. $p=0.75$; $m=1,2,3,\dots,10$

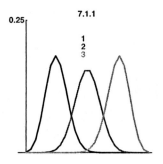

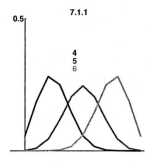

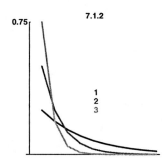

7.1.3   Hypergeometric density, $P(m|n,N,p) = \dfrac{\dbinom{Np}{m}\dbinom{N(1-p)}{n-m}}{\dbinom{N}{n}},$

where: m, number of items of a given type in a sample with the upper bound given by $m = \min(n,Np)$; n, sample size; N, total number of items available $(N > n)$; p, probability of a given item type in total number N.

1. $n=10$, $N=40$, $p=0.25$; $m=0,1,2,\ldots,10$
2. $n=10$, $N=40$, $p=0.50$; $m=0,1,2,\ldots,10$
3. $n=10$, $N=40$, $p=0.75$; $m=0,1,2,\ldots,10$

4. $n=20$, $N=40$, $p=0.25$; $m=0,1,2,\ldots,20$
5. $n=20$, $N=40$, $p=0.50$; $m=0,1,2,\ldots,20$
6. $n=20$, $N=40$, $p=0.75$; $m=0,1,2,\ldots,20$

7.1.4   Negative binomial density, $P(m|n,p) = \dbinom{n+m-1}{m}p^{n}(1-p)^{m}$

where: m, number of failures prior to nth success; n, number of successes; p, probability of a given event.

1. $n=10$, $p=0.25$; $m=0,1,2,\ldots,50$
2. $n=10$, $p=0.50$; $m=0,1,2,\ldots,50$
3. $n=10$, $p=0.75$; $m=0,1,2,\ldots,50$

4. $n=25$, $p=0.25$; $m=0,1,2,\ldots,100$
5. $n=25$, $p=0.50$; $m=0,1,2,\ldots,100$
6. $n=25$, $p=0.75$; $m=0,1,2,\ldots,100$

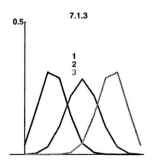

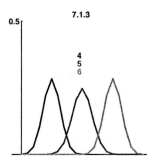

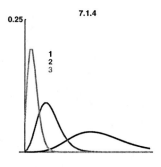

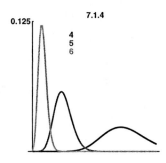

7.1.5   Poisson density, $P(m|r) = \dfrac{e^{-r} r^m}{m!}$

where: m, number of events occurring in a given unit of time; r, mean rate (number of events per unit time).

1. $r=2.0$; $m=0,1,2,...,25$
2. $r=6.0$; $m=0,1,2,...,25$
3. $r=10.0$; $m=0,1,2,...,25$

7.2 Continuous Probability Densities

The following probability densities are continuous functions, plotted such that the x axis limits are -1 to $+1$ (with the actual domain of x given by the argument as listed). The range of y is arbitrary, selected only to plot the function in an easily viewable manner. As for the discrete densities, the scale may change among plots for a given function. A property common to all continuous probability densities is that the integral equals unity; thus:

$$\int_a^b P(x)\mathrm{d}x = 1,$$

where a and b are the domain of the particular density function.

7.2.1   Beta density, $P(x) = \left[\dfrac{1}{B(a,b)}\right]^{a-1} (1-x)^{b-1},$

where B is the beta function (see Section 5.3.3).

Domain: $0<x<1$

1. $0.4\ P(x)$: $a=2$, $b=1$
2. $0.4\ P(x)$: $a=2$, $b=2$
3. $0.4\ P(x)$: $a=2$, $b=3$
4. $0.4\ P(x)$: $a=2$, $b=4$

5. $0.3\ P(x)$: $a=3$, $b=1$
6. $0.3\ P(x)$: $a=3$, $b=2$
7. $0.3\ P(x)$: $a=3$, $b=3$
8. $0.3\ P(x)$: $a=3$, $b=4$

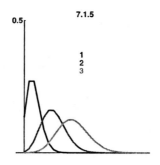

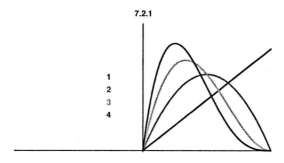

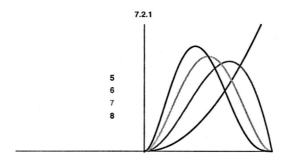

7.2.2   Cauchy density, $P(x) = \dfrac{1}{\pi b}\left[1 + \left(\dfrac{x-a}{b}\right)^2\right]^{-1}$

Domain: $-\infty < x < \infty$

   1. $2.0\ P(5x)$: $a=0$, $b=1$
   2. $2.0\ P(5x)$: $a=0$, $b=2$
   3. $2.0\ P(5x)$: $a=0$, $b=3$

7.2.3   Chi-Square density, $P(x) = \dfrac{1}{2^{n/2}\Gamma(n/2)}x^{(n-2)/2}e^{-x/2}$,

where Γ is the gamma function (see Section 5.3.1).

Domain: $0 < x < \infty$

   1. $5.0\ P(50x)$: $n=5$
   2. $5.0\ P(50x)$: $n=15$
   3. $5.0\ P(50x)$: $n=25$

7.2.4   Exponential density, $P(x) = \dfrac{1}{b}e^{-\frac{x-a}{b}}$

Domain: $a < x < \infty$

   1. $P(5x)$: $a=0$, $b=1$
   2. $P(5x)$: $a=0$, $b=2$
   3. $P(5x)$: $a=0$, $b=3$

7.2.5   Extreme value density, $P(x) = \dfrac{1}{b}e^{-\left|\frac{x-a}{b}\right| - \exp\left(-\left|\frac{x-a}{b}\right|\right)}$

Domain: $-\infty < x < \infty$

   1. $2.0\ P(5x)$: $a=0$, $b=1$
   2. $2.0\ P(5x)$: $a=0$, $b=2$
   3. $2.0\ P(5x)$: $a=0$, $b=3$

7.2.2

1
2
3

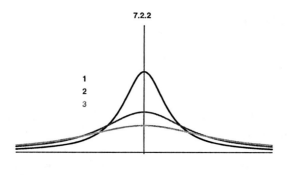

7.2.3

1
2
3

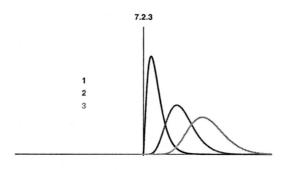

7.2.4

1
2
3

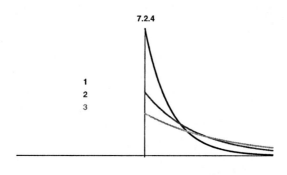

7.2.5

1
2
3

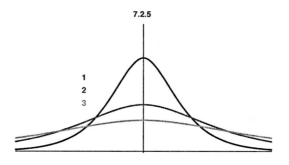

7.2.6   Gamma density, $P(x) = \dfrac{1}{\Gamma(a)b^a} x^{a-1} e^{-\frac{x}{b}} \, (a, b > 0)$,

and where Γ is the gamma function (see Section 5.3.1).

Domain: $x > 0$

1. $P(10x)$: $a = 2$, $b = 0.5$
2. $P(10x)$: $a = 2$, $b = 1.0$
3. $P(10x)$: $a = 2$, $b = 2.0$

4. $P(10x)$: $a = 3$, $b = 0.5$
5. $P(10x)$: $a = 3$, $b = 1.0$
6. $P(10x)$: $a = 3$, $b = 2.0$

7.2.7   Laplace density, $P(x) = \dfrac{1}{2b} e^{-\left| \frac{x-a}{b} \right|}$

Domain: $-\infty < x < \infty$

1. $2.0 \, P(5x)$: $a = 0$, $b = 1$
2. $2.0 \, P(5x)$: $a = 0$, $b = 2$
3. $2.0 \, P(5x)$: $a = 0$, $b = 3$

7.2.8   Logistic density, $P(x) = \dfrac{1}{b} e^{\frac{x-a}{b}} \left(1 + e^{\frac{x-a}{b}} \right)^{-2}$

Domain: $0 < x < \infty$

1. $4.0 \, P(10x)$: $a = 0$, $b = 1$
2. $4.0 \, P(10x)$: $a = 0$, $b = 2$
3. $4.0 \, P(10x)$: $a = 0$, $b = 3$

7.2.6

1
2
3

7.2.6

4
5
6

7.2.7

1
2
3

7.2.8

1
2
3

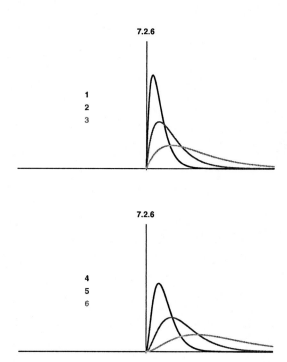

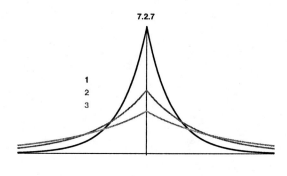

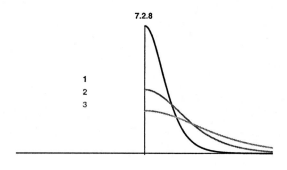

7.2.9   Log-normal density, $P(x) = \dfrac{1}{\sqrt{2\pi}b}\,\mathrm{e}^{-\frac{[(\ln x - a)/b]^2}{2}}$

Domain: $x > 0$

1. $P(10x)$: $a = 0$, $b = 0.5$
2. $P(10x)$: $a = 0$, $b = 1.0$
3. $P(10x)$: $a = 0$, $b = 2.0$

7.2.10   Maxwell density, $P(x) = \dfrac{4}{\sqrt{\pi}a^3}x^2\mathrm{e}^{-\left(\frac{x}{a}\right)^2}$

Domain: $x > 0$

1. $P(10x)$: $a = 1$
2. $P(10x)$: $a = 2$
3. $P(10x)$: $a = 3$

7.2.11   Normal (Gaussian) density, $P(x) = \dfrac{1}{\sqrt{2\pi}b}\,\mathrm{e}^{-\frac{[(x - a)/b]^2}{2}}$

Domain: $-\infty < x < \infty$

1. $2.0\,P(10x)$: $a = 0$, $b = 1$
2. $2.0\,P(10x)$: $a = 0$, $b = 2$
3. $2.0\,P(10x)$: $a = 0$, $b = 3$

7.2.12   Pareto density, $P(x) = \dfrac{x}{b\left(1 + \dfrac{x}{b}\right)^{a+1}}$

Domain: $x > 0$

1. $5.0\,P(10x)$: $a = 2$, $b = 1$
2. $5.0\,P(10x)$: $a = 2$, $b = 2$
3. $5.0\,P(10x)$: $a = 2$, $b = 3$

7.2.9

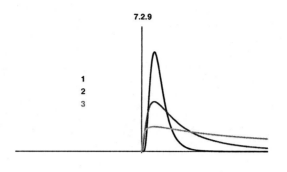

1
2
3

7.2.10

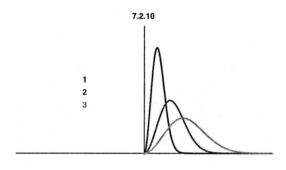

1
2
3

7.2.11

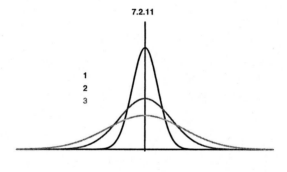

1
2
3

7.2.12

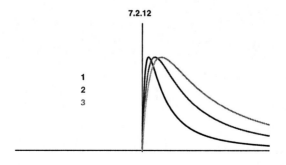

1
2
3

4. 10.0 $P(10x)$: $a=4$, $b=1$
5. 10.0 $P(10x)$: $a=4$, $b=2$
6. 10.0 $P(10x)$: $a=4$, $b=3$

7.2.13   Rayleigh density, $P(x) = \dfrac{1}{a^2} x e^{-\frac{|x/a|^2}{2}}$

Domain: $x>0$

1. $P(10x)$: $a=1$
2. $P(10x)$: $a=2$
3. $P(10x)$: $a=3$

7.2.14   Snedecor's F density, $P(x) = \dfrac{m^{m/2} n^{n/2}}{B(m/2, n/2)} \dfrac{x^{(m-2)/2}}{(n+mx)^{(m+n)/2}}$

where B is the beta function (see Section 5.3.3).

Domain: $x>0$

1. $P(5x)$: $m=5$, $n=10$
2. $P(5x)$: $m=15$, $n=10$
3. $P(5x)$: $m=50$, $n=10$

4. 0.5 $P(5x)$: $m=5$, $n=20$
5. 0.5 $P(5x)$: $m=15$, $n=20$
6. 0.5 $P(5x)$: $m=50$, $n=20$

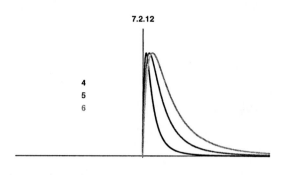

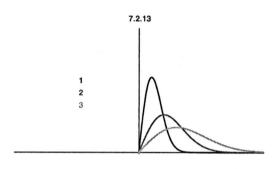

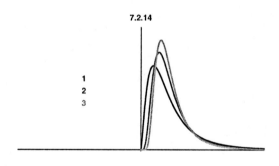

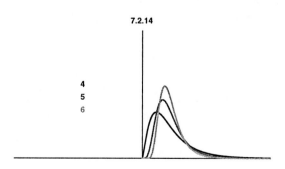

7.2.15   Student's t density, $P(x) = \dfrac{1}{\sqrt{n}B\left(\frac{1}{2},\frac{n}{2}\right)} \dfrac{1}{\left(1+\dfrac{x^2}{n}\right)^{(n+1)/2}}$,

where B is the beta function (see Section 5.3.3).

Domain: $-\infty < x < \infty$

    1. 2.0 $P(5x)$: $n=2$
    2. 2.0 $P(5x)$: $n=5$
    3. 2.0 $P(5x)$: $n=25$

7.2.16   Weibull density, $P(x) = \dfrac{b}{a^b}x^{b-1}e^{-\left(\frac{x}{a}\right)^b}$

Domain: $x > 0$

    1. 0.5 $P(5x)$: $a=1, b=1$
    2. 0.5 $P(5x)$: $a=1, b=2$
    3. 0.5 $P(5x)$: $a=1, b=3$

    4. 1.0 $P(5x)$: $a=2, b=2$
    5. 1.0 $P(5x)$: $a=2, b=3$
    6. 1.0 $P(5x)$: $a=2, b=4$

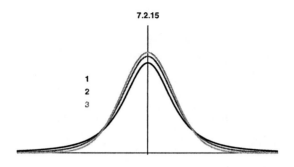

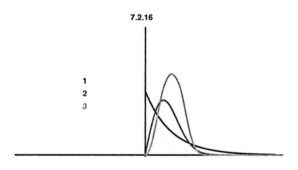

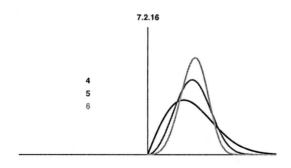

7.3 Sampling Distributions

The following sampling distributions are expressed as integrals of a density function. By definition, at the upper limit the integral equals unity; therefore, the distributions are plotted such that the maximum is always unity. The actual domain of the sampling variable is as listed.

7.3.1   Normal distribution, $P(X) = \dfrac{1}{(2\pi)^{1/2}b} \displaystyle\int_{-\infty}^{X} \exp\left\{\dfrac{[(x-a)/b]^2}{2}\right\} dx$

Note: $P(x)$ is expressible as $\dfrac{1}{2}\left[1 + \mathrm{Erf}\left(\dfrac{x-a}{2^{1/2}b}\right)\right]$ where Erf is the error function (see Section 5.4.1).

Domain: $-\infty < X < \infty$

    1. $P(5X)$: $a=0.0$, $b=0.5$
    2. $P(5X)$: $a=0.0$, $b=1.0$
    3. $P(5X)$: $a=0.0$, $b=2.0$

7.3.2   Student t distribution, $P(t|n) = \dfrac{\int_{-\infty}^{t}[1 + (x^2/n)]^{-(n+1)/2} dx}{n^{1/2}B\left(\frac{1}{2}, \frac{n}{2}\right)}$

where B is the beta function (see Section 5.3.3).

Domain: $-\infty < t < \infty$

    1. $P(5t)$: $n=2$
    2. $P(5t)$: $n=5$
    3. $P(5t)$: $n=99$

7.3.3   Chi-Square distribution, $P(\chi^2|n) = \dfrac{\int_{0}^{\chi^2} x^{(n-2)/2} e^{-x/2} dx}{2^{n/2}\Gamma(n/2)}$,

where Γ is the gamma function (see Section 5.3.1).

Domain: $\chi^2 > 0$

    1. $P(25\chi^2)$: $n=2$
    2. $P(25\chi^2)$: $n=6$
    3. $P(25\chi^2)$: $n=12$

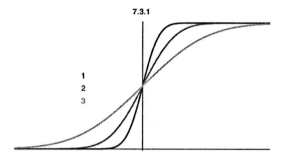

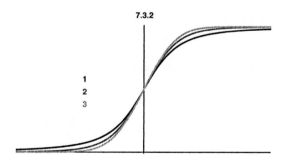

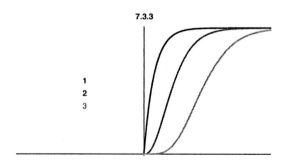

7.3.4 F distribution, $P(F|m,n) = \dfrac{m^{m/2}n^{n/2}}{B\left(\frac{m}{2},\frac{n}{2}\right)} \displaystyle\int_0^F x^{(m-2)/2}(n+mx)^{-(m+n)/2}\mathrm{d}x,$

and where B is the beta function (see Section 5.3.3).

Domain: $F > 0$

  1. $P(5F)$: $m = 2$, $n = 10$
  2. $P(5F)$: $m = 6$, $n = 10$
  3. $P(5F)$: $m = 20$, $n = 10$

  4. $P(5F)$: $m = 2$, $n = 20$
  5. $P(5F)$: $m = 6$, $n = 20$
  6. $P(5F)$: $m = 20$, $n = 20$

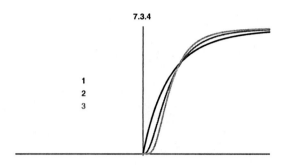

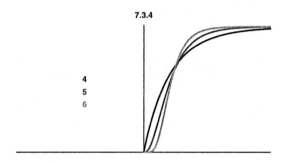

8

Nondifferentiable and Discontinuous Functions

In the equations of this chapter, the symbol H is used for the *unit step function* and the symbol δ for the *unit impulse function* (also called the *Dirac delta function*). The function δ is defined only over an infinitesimal interval of x such that its integral over that interval is unity. This requires δ to have an infinite amplitude, and the amplitude is truncated here at $y=1$ for purposes of illustration. The function H is defined such that $H(x-a)$ is zero for $x<a$ and $H(x-a)=1$ for $x \geq a$. Therefore, $H(x-a)$ is the integral of $\delta(x-a)$.

8.1   Functions with a Finite Number of Discontinuities

8.1.1 $y = \delta(x-a)$

(*Dirac*) *delta function*
Note:
$$\delta(x-a) = d[H(x-a)]/dx.$$
$a = 0.5$

8.1.2 $y = \delta'(x-a)$

Doublet function
$a = 0.5$

8.1.3 $y = c[H(x-a)]$

(*Heaviside*) *step function*
Note:
$$H(x-a) = [1 + \operatorname{sgn}(x-a)]/2, \quad \text{except at } x = a.$$
$a = 0.5, c = 0.5$

8.1.4 $y = c[H(x-a) - H(x-b)]$

Boxcar function
$a = 0.25, b = 0.75, c = 0.5$

8.1.1

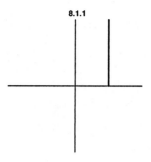

8.1.2

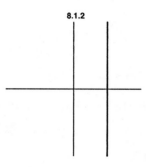

8.1.3

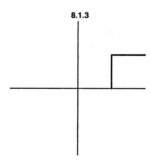

8.1.4

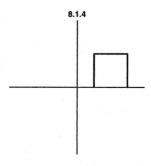

8.1.5 $y = c[H(x-a) - 2H(x-b) + H(x-2b+a)]$

Double boxcar function

$a=0.25, b=0.5, c=0.5$

8.2   Functions with an Infinite Number of Discontinuities

8.2.1 $y = c \sum_{n=0}^{\infty} H(x-na)$

Stairstep function

$a=0.2, c=0.1$

8.2.2 $y = \sum_{n=1}^{\infty} c^n H(x-na)$

   a. $a=0.2, c=0.5$

   b. $a=0.2, c=-0.5$

8.1.5

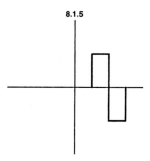

8.2.1

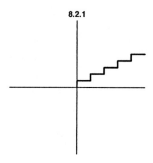

8.2.2a

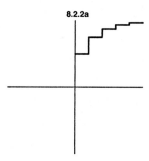

8.2.2b

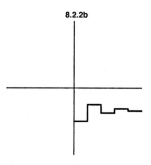

8.2.3 $y = c \; \text{sgn}[\sin(2\pi ax)]$

Square sine wave
$a = 5.0, \; c = 0.5$

8.2.4 $y = c \; \text{sgn}[\cos(2\pi ax)]$

Square cosine wave
$a = 5.0, \; c = 0.5$

8.2.5 $y = c \; \text{arctan}[\tan(2\pi ax)]$

Sawtooth wave
$a = 2.5, \; c = 0.35$

8.2.6 $y = c \; \text{arctan}[\tan(-2\pi ax)]$

Reverse sawtooth wave
$a = 2.5, \; c = 0.35$

8.2.3

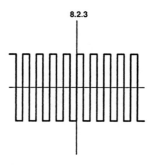

8.2.4

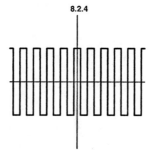

8.2.5

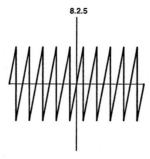

8.2.6

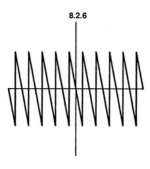

8.2.7 $y = c\{\text{sgn}[\sin(2\pi ax)] + 1\}$

Comb function

$a = 5.0, c = 0.5$

8.3 Functions with a Finite Number of Discontinuities in First Derivative

8.3.1 $y = [c/(b-a)][(x-a)H(x-a) - (x-b)H(x-b)]$

Ramp function
$a = -0.5, b = 0.5, c = 0.5$

8.3.2 $y = c(1 - |x|/a)[H(x+a) - H(x-a)]$

Triangular function

$a = 0.5, c = 0.5$

8.3.3 $y = c(1 - x^2/a^2)^{1/2}[H(x+a) - H(x-a)]$

Semiellipse (semicircle for $a = c$)

$a = 0.75, c = 0.5$

8.2.7

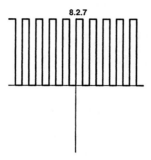

8.3.1

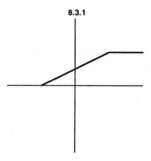

8.3.2

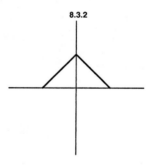

8.3.3

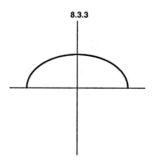

8.3.4 $y = c(1 - e^{-ax})H(x)$

Exponential ramp

$a = 5.0, c = 0.5$

8.4 Functions with an Infinite Number of Discontinuities in First Derivative

8.4.1 $y = c \arcsin[\sin(2\pi ax)]$

Triangular sine wave

$a = 5.0, c = 0.35$

8.4.2 $y = c \arccos[\cos(2\pi ax)]$

Triangular cosine wave

$a = 5.0, c = 0.35$

8.4.3 $y = c\{\arcsin[\sin(2\pi ax)] + \pi/2\}$

Rectified triangular sine wave

$a = 5.0, c = 0.2$

8.3.4

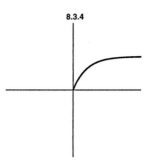

8.4.1

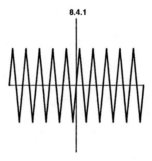

8.4.2

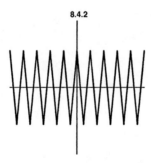

8.4.3

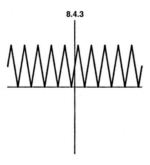

8.4.4 $y = c\{\arcsin[\cos(2\pi ax)] + \pi/2\}$

Rectified triangular cosine wave

$a = 5.0$, $c = 0.2$

8.4.5 $y = c|\sin(2\pi ax)|$

Rectified sine wave

$a = 2.5$, $c = 0.5$

8.4.6 $y = c|\cos(2\pi ax)|$

Rectified cosine wave

$a = 2.5$, $c = 0.5$

8.4.7 $y = \sum\limits_{k=1}^{\infty} \dfrac{\sin(\pi k^n x)}{\pi k^n}$

Weierstrass function

1. $2.0\,y$; $n = 2$; $0 < x < 1$
2. $2.0\,y$; $n = 3$; $0 < x < 1$
3. $2.0\,y$; $n = 4$; $0 < x < 1$

8.4.4

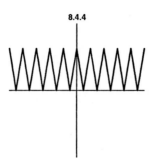

8.4.5

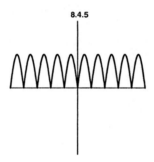

8.4.6

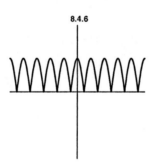

8.4.7

1
2
3

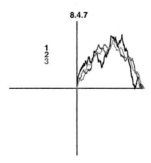

9

Random Processes

A large number of classes of random processes exist, and many variations are recognized within each class. Extensive treatments from differing perspectives of random processes can be found in Mandelbrot,[1] Box and Jenkins,[2] Parzen,[3] and Bendat and Piersol.[4] This chapter only attempts to show those one-dimensional processes that are of simple form or are in common usage. There is a large and complex suite of random processes called *Markov processes*; they are of such variety that no representative examples are given here.

The random processes of this chapter are plotted as time evolutions, with the time axis being horizontal. For each process, three realizations are shown, each independent of the others, so that the reader can appreciate the variability of each process. For each realization, 200 (sometimes 100 or 50) points or increments are plotted. The plots are constructed with arbitrary scaling of the amplitude, but such that the plotted range of the amplitude is always constant for all three realizations in a given process.

9.1 Elementary Random Processes

9.1.1 White Noise

$$y(i) = r(i),$$

where r is taken from a Gaussian density (see Section 7.2.11) with zero mean and unit standard deviation.

9.1.2 Two-Valued Process

$$y(i) = (1|-1),$$

where the random variable $+1$ or -1 is chosen with equal probability at each step.

9.1.3 Unit Random Walk

$$y(i) = \sum_{n=0}^{i} (1|-1),$$

where the sum (integral for continuous time) is of the two-valued process of 9.1.2.

9.2 General Linear Processes

The discrete, general linear mixed (ARMA) process is described by the equation:

$$y(i) = a_1 y(i-1) + a_2 y(i-2) + \cdots + a_n y(i-n) + b_1 r(i-1) + b_2 r(i-2) + \cdots + b_m r(i-m) r(i),$$

where the a_j and b_j are constant coefficients and r is a random variable. If all $b_j=0$, the process is called an *autoregressive (AR) process* of order n. If all $a_j=0$, the process is called a *moving-average (MA) process* of order m. If at least one a_j and at least one b_j are non-zero, the process is called a *mixed (ARMA) process*.

9.1.1

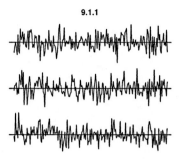

9.1.2

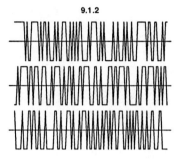

9.1.3

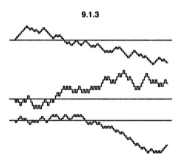

9.2.1 First-Order Autoregressive

a. $a_1 = 0.2$

b. $a_1 = 0.8$

c. $a_1 = -0.2$

d. $a_1 = -0.8$

9.2.1a

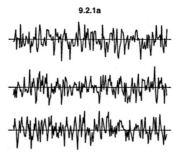

9.2.1b

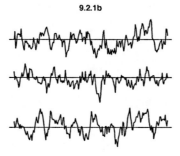

9.2.1c

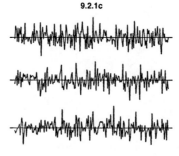

9.2.1d

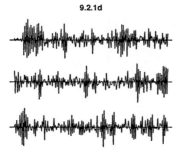

9.2.2 Second-Order Autoregressive

a. $a_1 = 0.4$, $a_2 = 0.4$

b. $a_1 = -0.4$, $a_2 = 0.4$

c. $a_1 = 0.4$, $a_2 = -0.4$

d. $a_1 = -0.4$, $a_2 = -0.4$

9.2.2a

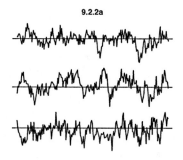

9.2.2b

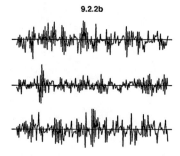

9.2.2c

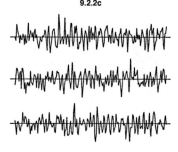

9.2.2d

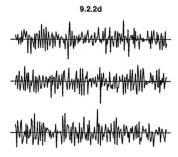

9.2.3 First-Order Moving Average

a. $b_1 = 0.2$

b. $b_1 = 0.8$

c. $b_1 = -0.2$

d. $b_1 = -0.8$

9.2.3a

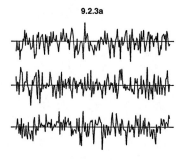

9.2.3b

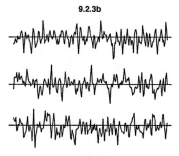

9.2.3c

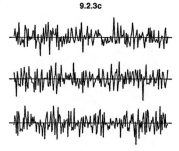

9.2.3d

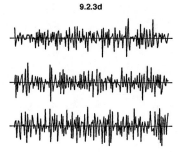

9.2.4   Second-Order Moving Average

a. $b_1 = 0.4$, $b_2 = 0.4$

b. $b_1 = -0.4$, $b_2 = 0.4$

c. $b_1 = 0.4$, $b_2 = -0.4$

d. $b_1 = -0.4$, $b_2 = -0.4$

9.2.4a

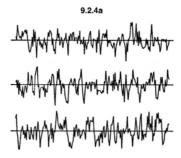

9.2.4b

9.2.4c

9.2.4d

9.2.5 First-Order Mixed

a. $a_1 = 0.4$, $b_1 = 0.4$

b. $a_1 = -0.4$, $b_1 = 0.4$

c. $a_1 = 0.4$, $b_1 = -0.4$

d. $a_1 = -0.4$, $b_1 = -0.4$

9.2.5a

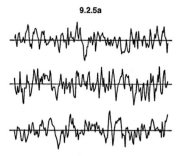

9.2.5b

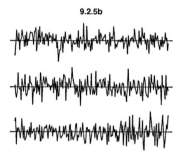

9.2.5c

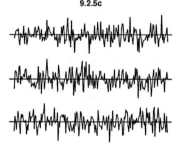

9.2.5d

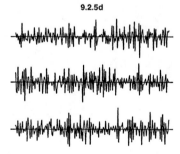

9.2.6 High-Order Moving Average

Set all $a_i=0$ and all $b_i=1$ in the ARMA process

    a. $m=4$

    b. $m=9$

    c. $m=16$

    d. $m=25$

9.3 Integrated Processes

The discrete, general *integrated mixed (ARIMA) process*, $y(i)$, is described by the equation:

$$\nabla^d y(i) = a_1 \nabla^d y(i-1) + a_2 \nabla^d y(i-2) + \cdots + a_n \nabla^d y(i-n) + b_1 r(i-1) + b_2 r(i-2) + \cdots$$
$$+ b_m r(i-m) + r(i),$$

9.2.6a

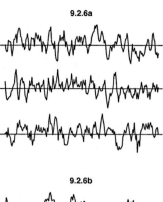

9.2.6b

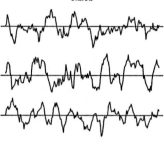

9.2.6c

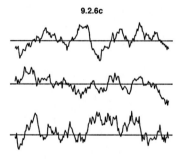

9.2.6d

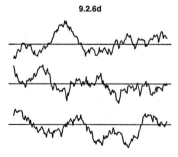

where the a_j and b_j are constant coefficients, r is a Gaussian random variable, and d is the order of the differential. For example, if $d=1$, then y is the first integral of an ordinary ARMA process (see the explanation of a general linear process in Section 9.2.) If all $b_j=0$, the process is called an *integrated autoregressive (ARI) process* of order n. If all $a_j=0$, the process is called an *integrated moving-average (IMA) process* of order m. If at least one a_j and at least one b_j are non-zero, the process is an integrated mixed (ARIMA) process.

9.3.1 First-Order Autoregressive

a. $a_1=0.2$

b. $a_1=0.8$

c. $a_1=-0.2$

d. $a_1=-0.8$

9.3.1a

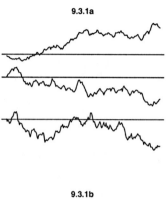

9.3.1b

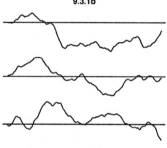

9.3.1c

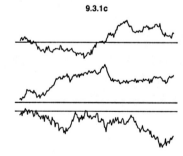

9.3.1d

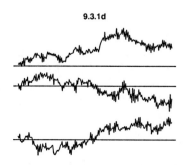

9.3.2 First-Order Moving Average

a. $b_1 = 0.2$

b. $b_1 = 0.8$

c. $b_1 = -0.2$

d. $b_1 = -0.8$

9.3.2a

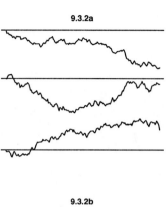

9.3.2b

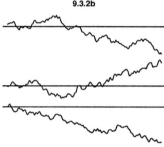

9.3.2c

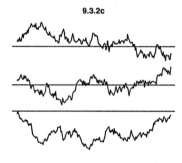

9.3.2d

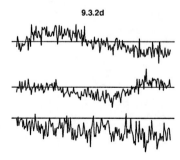

9.3.3 First-Order Mixed

a. $a_1 = 0.4$, $b_1 = 0.4$

b. $a_1 = -0.4$, $b_1 = 0.4$

c. $a_1 = 0.4$, $b_1 = -0.4$

d. $a_1 = -0.4$, $b_1 = -0.4$

9.4 Fractal Processes

The general *fractal process* (fractal line-to-line function) is a process whose increments are distributed according to the Gaussian (or normal) density law. If $y(t - \Delta t)$ and $y(t + \Delta t)$ are

9.3.3a

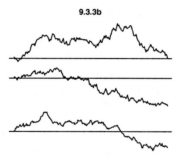

9.3.3b

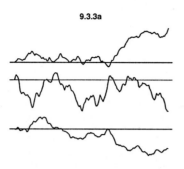

9.3.3c

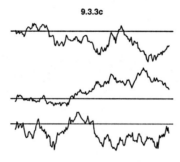

9.3.3d

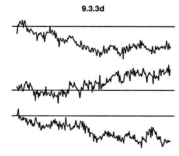

two values separated by a constant increment then the correlation of the two values is given by:

$$r = 2^{2H-1} - 1.$$

When $H=1/2$, $r=0$ and the process is the classical one-dimensional *Brownian motion* because each new value is incremented by an independent random variable from the last value. Processes for which $H>1/2$ ($r>0$) are called *persistent fractal processes* because they have long wavelength components, and processes for which $H<1/2$ ($r<0$) are called *anti-persistent fractal processes* because they are dominated by short wavelengths. H is in the range of 0 to 1; therefore, r is in the range of $-1/2$ to 1.

The general fractal process is also called a *fractional Brownian process*. The method of construction is an approximate method taken from K. Falconer[5] (Equation 16.13).

9.4.1   Brown Function

The Brown function is also called a *Bachelier* or *Wiener* or *Levy function*.

$$H = 0.5$$

9.4.2   Persistent Fractal Process

$$H = 0.8$$

9.4.3   Antipersistent Fractal Process

$$H = 0.2$$

9.4.1

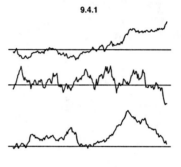

9.4.2

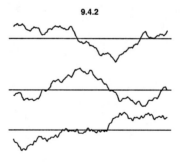

9.4.3

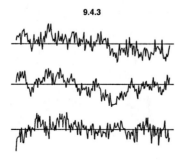

9.5   Poisson Processes

In the *Poisson process*, events occur over time at a mean rate of a. Let $N(t)$ be the number of events which have occurred since $t = 0$. Then the probability that $N(t) = m$ is given by (see Section 7.1.5):

$$P[N(t) = m] = e^{-at}(at)^m/m$$

The interevent time, T, has an exponential probability density; thus,

$$P(T) = ae^{-aT}.$$

The above is called the ordinary or *homogeneous Poisson process*. When the mean rate, a, of events varies with time, it is a *nonhomogeneous Poisson process*. Another variation is to not track the cumulative number of events but rather to track the accumulation of a random variable at the times given by an ordinary Poisson process; such processes are called *compound Poisson processes*.

9.5.1   Homogeneous Poisson Process

$$a = 1.0$$

9.5.2   Nonhomogeneous Poisson Process, with $a = a_0 + a_1 t$

$$a_0 = 1.0, \quad a_1 = 0.05$$

9.5.3   Compound Poisson Process

The random variable is from a normal distribution with mean zero.

$$a = 1.0$$

9.5.1

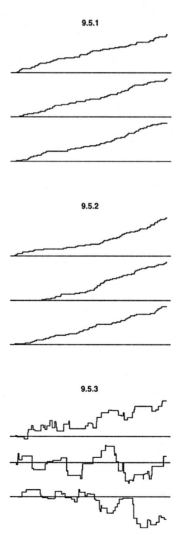

9.5.2

9.5.3

9.5.4 Poisson Wave (Telegraph Signal)

This process takes one of the values -1 or $+1$, with equal probability, at times corresponding to a Poisson process.

$$a = 1.0$$

References

1. B.B. Mandelbrot. 1983. *The Gractal Geometry of Nature*, San Francisco: W.H., Freeman.
2. G.E.P. Box, G.M. Jenkins, and G.C. Reinsel. 1994. *Time Series Analysis: Forecasting and Control*, 3rd Ed., Englewood Cliffs, London: Prentice Hall.
3. E. Parzen. 1962. *Stochastic Processes, Classics in Applied Mathematics*, 24, Philadelphia: Society of Industrial and Applied Mathematics.
4. J.S. Bendat and A.G. Piersol. 2000. *Random Data: Analysis and Measurement Procedures*, New York: Wiley.
5. K. Falconer. 1990. *Fractal geometry: Mathematical Foundations and Applications*, New York: Wiley.

9.5.4

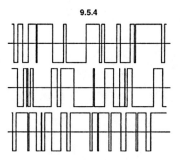

10

Polygons

The familiar shapes of two-dimensional geometry are shown in the first sections of this chapter. Scale is entirely relative for these figures. Later sections show how triangles, squares, and hexagons can be combined into more complicated shapes; these can serve as building blocks for even larger patterns, and some are capable of tiling the plane. The final section treats closed curves that are not strictly polygons because they have arc-like edges rather than straight-line edges.

10.1   Regular Polygons

1. $n=3$ sides (*equilateral triangle*)
2. $n=4$ sides (*square*)
3. $n=5$ sides (*pentagon*)
4. $n=6$ sides (*hexagon*)
5. $n=7$ sides (*heptagon*)
6. $n=8$ sides (*octagon*)
7. $n=9$ sides (*nonagon*)
8. $n=10$ sides (*decagon*)
9. $n=11$ sides (*undecagon*)
10. $n=12$ sides (*dodecagon*)

10.2   Star Polygons

1. $n=3$ points
2. $n=4$ points
3. $n=5$ points
4. $n=6$ points
5. $n=7$ points
6. $n=8$ points
7. $n=9$ points
8. $n=10$ points
9. $n=11$ points
10. $n=12$ points

10.3   Irregular Triangles

10.3.1   Right Triangle (One Angle $=90°$)

10.1.0

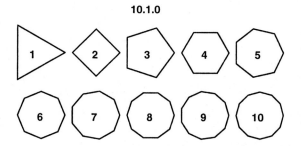

10.2.0

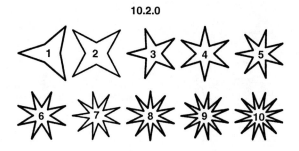

10.3.1

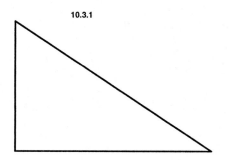

10.3.2   Isosceles Triangle (Two Angles Equal)

10.3.3   Acute Triangle (All Angles < 90°)

10.3.4   Obtuse Triangle (One Angle > 90°)

10.4   Irregular Quadrilaterals

For the following figures, let a, b, c, d be the length of the four sides.

10.4.1   Rectangle ($a=b$ and $c=d$; All Angles$=90°$)

10.3.2

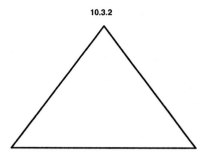

10.3.3

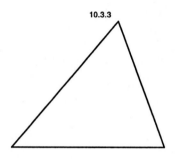

10.3.4

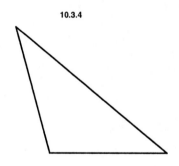

10.4.1

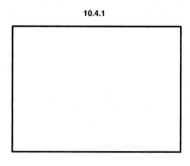

10.4.2 Parallelogram   ($a=b$ and $c=d$; All Angles $\neq 90°$)

10.4.3 Rhombus   ($a=b=c=d$; All Angles $\neq 90°$)

10.4.4 Trapezoid   ($a=b$; c and d Parallel; All Angles $\neq 90°$)

10.4.5 Deltoid   ($a=b$; and $c=d$; Two Angles Are Equal)

Given a, c, and the angle θ between sides a and b, the figure is completely specified. A negative c will create a concave variation.

   a. $a=1.0, c=1.5, \theta=100°$

10.4.2

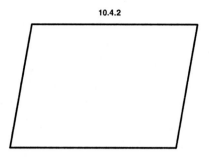

10.4.3

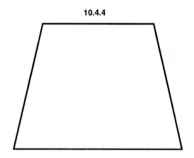

10.4.4

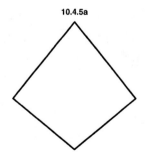

10.4.5a

b. $a = 1.0$, $c = -1.5$, $\theta = 100°$

10.5 Polyiamonds

10.5.1 Triamonds   (3 Connected Equilateral Triangles)

10.5.2 Tetriamonds   (4 Connected Equilateral Triangles)

10.5.3 Pentiamonds   (5 Connected Equilateral Triangles)

10.5.4 Hexiamonds   (6 Connected Equilateral Triangles)

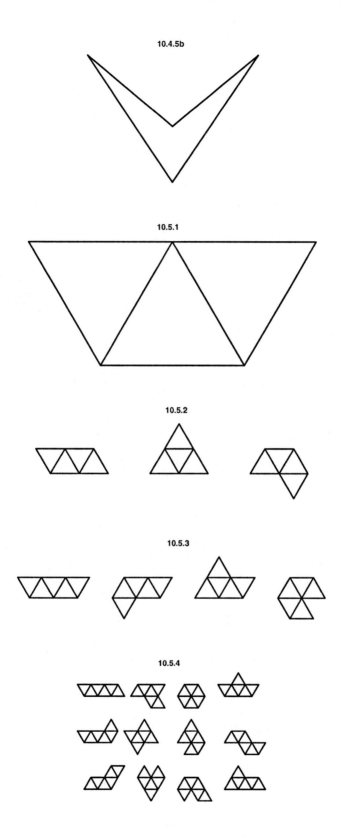

10.6 Polyominoes

10.6.1 Trominoes   (3 Connected Squares)

10.6.2 Tetrominoes   (4 Connected Squares)

10.6.3 Pentominoes   (5 Connected Squares)

10.7 Polyhexes

10.7.1 Trihexes   (3 Connected Regular Hexagons)

10.7.2 Tetrahexes   (4 Connected Regular Hexagons)

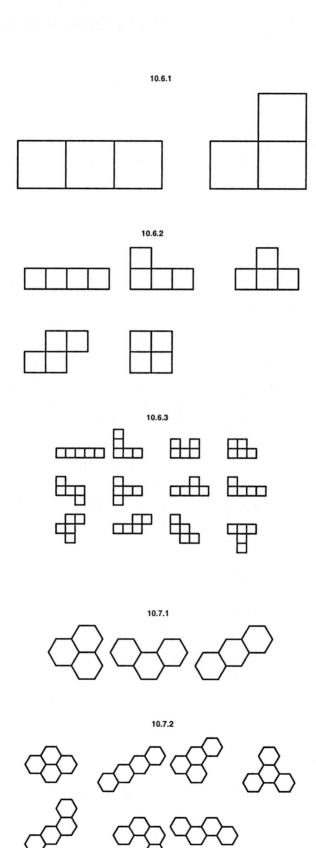

10.6.1

10.6.2

10.6.3

10.7.1

10.7.2

10.8 Miscellaneous Polygons

10.8.1 Reuleaux Polygon

For a *Reuleaux polygon*, the number of sides is always odd. The figure is made by centering a circle on a vertex of the regular polygon of n sides, letting its radius be the length of the sides of the polygon, and then using the arc between the two opposing points as a side of the Reuleaux polygon. Repeating for each of the vertices produces n connected arcs.

1. $n=3$ (*Reuleaux triangle*)
2. $n=5$ (*Reuleaux pentagon*)
3. $n=7$ (*Reuleaux heptagon*)
4. $n=9$ (*Reuleaux nonagon*)

10.8.2 Lune

A *lune* is constructed from two intersecting circles such that one edge is concave and the other is convex. Let the radii of the circles be a and b and the separation of their centers be c, with the circle of radius b lying to the right. Although two lunes are actually produced, these figures are of the left-hand lune.

a. $a=1.0, b=1.5, c=1.0$

b. $a=1.0, b=1.5, c=2.0$

c. $a=1.0, b=0.5, c=1.0$

10.8.1

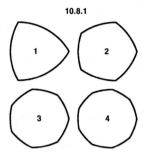

10.8.2a

10.8.2b

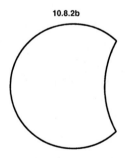

10.8.2c

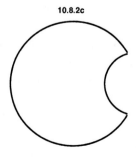

d. $a=1.0$, $b=0.5$, $c=0.5$

10.8.3 Lens

A *lens* is constructed from two intersecting circles, similar to the lune, but such that the figure with two convex edges is plotted. Let the radii of the circles be a and b and the separation of their centers be c, with the circle of radius b lying to the right.

a. $a=1.0$, $b=1.0$, $c=1.0$ (*symmetrical lens*)

b. $a=1.0$, $b=1.3$, $c=1.0$ (*asymmetrical lens*)

10.8.4 Annulus

The *annulus* is the area between two concentric circles of radii a and b.

$a=1.3$, $b=1.0$

10.8.5 Yin-Yang

The *yin-yang* figure is composed of three circles. The outer circle has radius a, and the two inner circles have radius $a/2$. The two inner circles are adjacent, with centers on the diameter of the larger circle. The outer circle is divided into semicircles of opposite colors. Each of the smaller circles also have halves of opposite color, arranged such that two areas of contiguous color appear within the area of the largest circle.

$a=1.0$

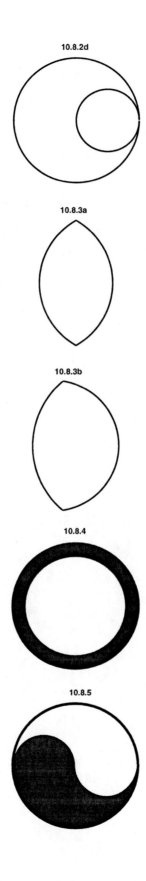

10.8.2d

10.8.3a

10.8.3b

10.8.4

10.8.5

11

Three-Dimensional Curves

As opposed to curves that lie wholly in a plane (called *plane curves*), those curves that occupy three dimensions are called *skew curves*. All three-dimensional curves must necessarily be expressed in parametric form:

$$x = f(t)$$
$$y = g(t)$$
$$z = h(t).$$

Because there are innumerable variations of the functions f, g, and h, three-dimensional curves can assume a wide variety of appearances. Only those curves having some accepted significance and use are illustrated here. Many interesting and useful three-dimensional curves can be generated simply by adding a z variation to the curves given in the previous chapters, after they are put into parametric form.

The curves in this chapter are plotted as points (x_p, y_p) projected on a plane that is normal to the vector between the origin $(0,0,0)$ and the viewpoint. The projection used is the *perspective projection* (see Foley and VanDam[1] for a full treatment of projections). If the viewing point is at large distance relative to the projected points (as is the case in this chapter), then the view approaches the *parallel projection* that is given by the transformations:

$$x_p = -x \sin \theta + y \cos \theta$$
$$y_p = -x \cos \theta \cos \phi - y \sin \theta \cos \phi + z \sin \phi$$

where (x,y,z) are the coordinates of the point on the curve prior to projection and (θ,ϕ) are the angles in spherical coordinates (see Section 1.3) of the vector normal to the projection plane. The three axes are plotted with solid lines between the limits of -1 and $+1$, with the positive z axis up.

11.1   Helical Curves

11.1.1   Circular Helix

Also called *right helicoid*.

$$x = a \sin t$$
$$y = a \cos t$$
$$z = t/(2\pi m)$$

$a = 0.5$, $n = 5$; $0 < t < 10\pi$; viewpoint $= (40, -50, 20)$

11.1.2   Elliptical Helix

$$x = a \sin t$$
$$y = b \cos t$$
$$z = t/(2\pi n)$$

$a = 0.3$, $b = 1.0$, $n = 5$; $0 < t < 10\pi$; viewpoint $= (40, -50, 20)$

11.1.3   Conical Helix

$$x = [at/(2\pi n)]\sin t$$
$$y = [at/(2\pi n)]\cos t$$
$$z = t/(2\pi n)$$

$a = 0.5$, $n = 5$; $0 < t < 10\pi$; viewpoint $= (40, -50, 20)$

11.1.4   Spherical Helix

$$x = \sin[t/(2n)]\cos t$$
$$y = \sin[t/(2n)]\sin t$$
$$z = \cos[t/(2n)]$$

$n = 5$; $0 < t < 10\pi$; viewpoint $= (40, -50, 20)$

11.1.1

11.1.2

11.1.3

11.1.4

11.1.5 N-Helix

Let n be the number of strands in the helix and m be the number of turns of each strand.

$$x = a \cos(t + 2\pi i/n) \quad i = 1, \ldots, n$$
$$y = a \sin(t + 2\pi i/n) \quad i = 1, \ldots, n$$
$$z = t/(2\pi m)$$

$a = 0.3$, $c = 3.0$, $n = 2$, $m = 3$; $0 < t < 6\pi$; viewpoint $= (40, -50, 20)$

11.2   Sine Waves in Three Dimensions

11.2.1 Sine Wave on Cylinder

$$x = a \cos t$$
$$y = a \sin t$$
$$z = c \cos(nt)$$

$a = 1.0$, $c = 0.5$, $n = 10$; $0 < t < 2\pi$; viewpoint $= (40, -50, 20)$

11.2.2 Sine Wave on Sphere

$$x = [b^2 - c^2 \cos^2(nt)]^{1/2} \cos t$$
$$y = [b^2 - c^2 \cos^2(nt)]^{1/2} \sin t$$
$$z = c \cos(nt)$$

$b = 1.0$, $c = 0.3$, $n = 10$; $0 < t < 2\pi$; viewpoint $= (40, -50, 20)$

11.2.3 Sine Wave on Hyperboloid of One Sheet

$$x = [b^2 + c^2 \cos^2(nt)]^{1/2} \cos t$$
$$y = [b^2 + c^2 \cos^2(nt)]^{1/2} \sin t$$
$$z = c \cos(nt)$$

$b = 1.0$, $c = 0.3$, $n = 10$; $0 < t < 2\pi$; viewpoint $= (40, -50, 20)$

11.1.5

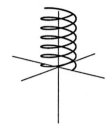

11.2.1

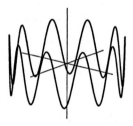

11.2.2

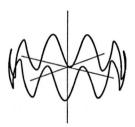

11.2.3

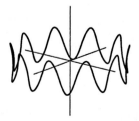

11.2.4 Sine Wave on Cone

$$x = a[1 + \cos(nt)]\cos t$$
$$y = a[1 + \cos(nt)]\sin t$$
$$z = c[1 + \cos(nt)]$$

$a=0.5$, $c=0.4$, $n=10$; $0<t<2\pi$; viewpoint$=(40,-50,20)$

11.2.5 Rotating Sine Wave

$$x = \sin(at)\cos(bt)$$
$$y = \sin(at)\sin(bt)$$
$$z = ct/(2\pi)$$

a. $a=3.0$, $b=1.0$, $c=1.0$; $-2\pi<t<2\pi$; viewpoint$=(40,-50,20)$

b. $a=3.0$, $b=0.3$, $c=1.0$; $-2\pi<t<2\pi$; viewpoint$=(40,-50,20)$

11.3 Miscellaneous 3-D Curves

11.3.1 Sici Spiral

$$x = a\,\mathrm{Ci}(t)$$
$$y = a\,\mathrm{Si}(t)$$
$$z = t/c,$$

where Si and Ci are the sine and cosine integrals (see Section 5.2).

$a=0.5$, $c=20.0$; $0.2<t<20$; viewpoint$=(40,-50,20)$

11.2.4

11.2.5a

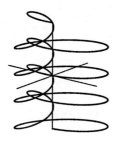

11.2.5b

11.3.1

11.3.2 Fresnel Integral Spiral

Also called Cornu's spiral.

$$x = C(t)$$
$$y = S(t)$$
$$z = t/c,$$

where S and C are the first and second Fresnel integrals (see Section 5.5).

$c=5.0$; $0<t<5$; viewpoint$=(40,-50,20)$

11.3.3 Toroidal Spiral

$$x = [a \sin(nt) + b]\cos t$$
$$y = [a \sin(nt) + b]\sin t$$
$$z = a \cos(nt)$$

$a=0.2$, $b=0.8$, $n=20$; $0<t<2\pi$; viewpoint$=(40,-50,20)$

11.3.4 Viviani's Curve

This curve is the intersection of a sphere and a cylinder.

$$x = [1 + \cos(t)]/2$$
$$y = \sin(t)/2$$
$$z = \sin(t/2)$$

$0<t<2\pi$; viewpoint$=(40,-50,20)$

11.3.5 Baseball Seam

The following was taken from a particular reference,[2] but it is not a unique parameterization of the baseball seam. Let the ball have radius R. An additional, arbitrary parameter is S, the arc distance between the closest points of the two seams. The seams touch when $S=0$. Baseballs are manufactured with roughly $S/R=0.818$.

Define:

$$x_0 = R \cos[S/(2R)]$$
$$y_0 = R \sin[S/(2R)]$$
$$b_0 = R/\sqrt{2}.$$

One-quarter of the figure, in the positive (y,z) quadrant, is then parameterized for $-x_0<t<x_0$; thus,

11.3.2

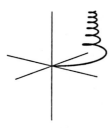

11.3.3

11.3.4

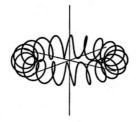

$$x = t$$

$$
y = \left\{
\begin{array}{l}
b_0 + \dfrac{(y_0 - b_0)|t|}{x_0} \, ; t > 0 \\[4mm]
\sqrt{R^2 - t^2 - \left[b_0 + \dfrac{(y_0 - b_0)|t|}{x_0} \right]^2} \; ; t < 0
\end{array}
\right\}
$$

$$z = \sqrt{R^2 - t^2 - y^2}.$$

For the second quarter ($y < 0$), reverse the sign of the y component. For the bottom half, reflect the two quarters about the x–y plane.

    a. $R = 1.0$, $S = 0.818$, $-x_0 < t < x_0$; viewpoint $= (40, -40, 8)$

    b. $R = 1.0$, $S = 0.0$, $-x_0 < t < x_0$; viewpoint $= (-40, -20, 8)$

11.3.6   3-D Astroid

$$x = \cos^3 t$$
$$y = \sin^3 t$$
$$z = \cos(2t)$$

$0 < t < 2\pi$; viewpoint $= (30, -50, 20)$

11.3.5a

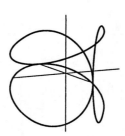

11.3.5b

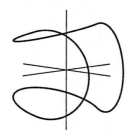

11.3.6

11.3.7 Spherical Spiral

This curve is the locus of a point as it turns around a sphere, descending from the top to the antipode. It is also called a *loxodrome* or *rhumb line*.

$$x = \cos t \, \cos(nt)$$

$$y = \sin t \, \sin(nt)$$

$$z = \sin t$$

$n = 20$; $-\pi/2 < t < \pi/2$; viewpoint $= (30, -50, 20)$

11.3.8 Bicylinder Curve

This curve is the intersection of two orthogonal cylinders of radii a and b.

$$x = a \cos t$$

$$y = a \sin t$$

$$z = [b^2 - a^2 \sin^2 t]^{1/2}$$

$a = 0.7$, $b = 1.0$; $0 < t < 2\pi$; viewpoint $= (30, -50, 20)$

11.4 Knots

A *knot* is a closed loop (no ends) such that it cannot be deformed into an *unknot*, which is a single open loop. The knots of this section are plotted with a small-diameter tube surrounding the trace of the knot curve to provide a clearer visual presentation of the crossing points. The viewpoint in all cases is from directly above the knot (on the positive z axis) at such a distance that the perspective is effectively the parallel one. The parametric representations used here are not necessarily unique; they merely provide the correct number and orientation of crossings. Some knots have lefthand and righthand versions; in such cases, the lefthand version has the same equations except for a change in sign for the x component of the parametric representation.

11.4.1 Unknot

$$x = \cos u$$

$$y = \sin u$$

$$z = 0$$

$0 < u < 2\pi$; viewpoint $= (0, 0, 10)$

11.3.7

11.3.8

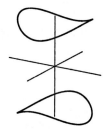

11.4.1

11.4.2 Trefoil Knot

The trefoil knot can be presented in either of two topologically equivalent forms. The first two figures have the parametric representation:

$$x = \pm\cos(pu)[r + \cos(qu)]$$
$$y = \sin(pu)[r + \cos(qu)]$$
$$z = \sin(qu).$$

The second two figures have the parametric representation:

$$x = -\sin(u)$$
$$y = \sin(3u/2) - \sin(u/2)/2$$
$$z = -\cos(3u/2).$$

a. Righthand version; $p=2$, $q=3$, $r=3$; $0<u<2\pi$; viewpoint $=(0,0,10)$

b. Lefthand version; $p=2$, $q=3$, $r=3$; $0<u<2\pi$; viewpoint $=(0,0,10)$

c. Righthand version; $p=2$, $q=3$, $r=3$; $0<u<4\pi$; viewpoint $=(0,0,10)$

d. Lefthand version; $p=2$, $q=3$, $r=3$; $0<u<4\pi$; viewpoint $=(0,0,10)$

11.4.2a

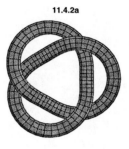

11.4.2b

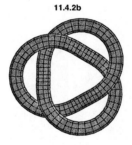

11.4.2c

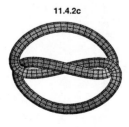

11.4.2d

11.4.3 Cinquefoil Knot (Solomon's Seal Knot)

Note that the parametric form is identical to that of the trefoil knot, except for a change of value for the parameter q.

$$x = \cos(pu)[r + \cos(qu)]$$
$$y = \sin(pu)[r + \cos(qu)]$$
$$z = \sin(qu)$$

    a. Righthand version; $p=2$, $q=5$, $r=3$; $0<u<2\pi$; viewpoint$=(0,0,10)$

    b. Lefthand version; $p=2$, $q=5$, $r=3$; $0<u<2\pi$; viewpoint$=(0,0,10)$

11.4.4 Figure Eight Knot

The figure eight knot can be presented in either of two topologically equivalent forms. The first two figures have the parametric representation:

$0<u<2\pi$	$2\pi<u<4\pi$
$x = \sin u - \sin(2u)$	$x = -\sin(u-2\pi)$
$y = \sin(3u/2)$	$y = -2\sin[(u-2\pi)/2]$
$z = \cos(5u/2)$	$z = -\cos[3(u-2\pi)/2]$

The second two figures have the parametric representation:

$$0<u<4\pi$$
$$x = \sin(3u/2) + \sin(u/2)/2$$
$$y = \cos(u/2) + \cos(3u/2)$$
$$z = \sin u + \sin(2u)$$

    a. Righthand version; $0<u<4\pi$; viewpoint$=(0,0,10)$

11.4.3a

11.4.3b

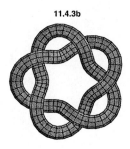

11.4.4a

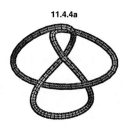

b. Lefthand version; $0 < u < 4\pi$; viewpoint = (0,0,10)

c. Righthand version; $0 < u < 4\pi$; viewpoint = (0,0,10)

d. Lefthand version; $0 < u < 4\pi$; viewpoint = (0,0,10)

11.4.5 Square Knot

$$x = \sin u + \sin(3u/2)$$
$$y = \cos(3u)$$
$$z = \cos(5u)$$

$0 < u < 2\pi$; viewpoint = (0,0,10)

11.4.4b

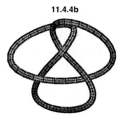

11.4.4c

11.4.4d

11.4.5

11.4.6 Granny Knot

$$x = \sin u + \sin(3u)/2$$
$$y = \cos(3u)$$
$$z = -\sin(6u) - \sin(8u)/2$$

a. Righthand version; $0 < u < 2\pi$; viewpoint = (0,0,10)

b. Lefthand version; $0 < u < 2\pi$; viewpoint = (0,0,10)

11.4.7 Miller Institute Knot

$$x = \sin(3u/2)$$
$$y = \cos u + \cos(u/2)/2$$
$$z = \sin(3u) + \sin(5u/2)/2$$

a. Righthand version; $0 < u < 4\pi$; viewpoint = (0,0,10)

b. Lefthand version; $0 < u < 4\pi$; viewpoint = (0,0,10)

11.4.6a

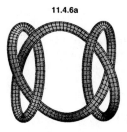

11.4.6b

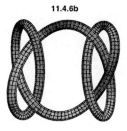

11.4.7a

11.4.7b

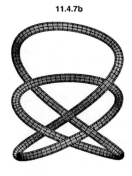

11.4.8 Lissajous Knot

A 3-D Lissajous knot is parameterized by:

$$x = \cos(n_x u + \phi_x)$$
$$y = \cos(n_y u + \phi_y)$$
$$z = \cos(n_z u + \phi_z).$$

A necessary, but not sufficient, condition for a Lissajous curve to form a knot is that n_x, n_y, and n_z are relatively prime; that is, no pair of these three numbers is divisible by a given integer. The phases (ϕ_x, ϕ_y, ϕ_z) are arbitrary.

    a. $n_x=2$, $n_y=3$, $n_z=5$, $\phi_x=\pi/12$, $\phi_y=\pi/2$, $\phi_z=0$; $0<u<2\pi$; viewpoint=(0,0,10)

    b. $n_x=2$, $n_y=3$, $n_z=7$, $\phi_x=\pi/12$, $\phi_y=\pi/2$, $\phi_z=0$; $0<u<2\pi$; viewpoint=(0,0,10)

11.5 Links

A *link* is formed by two or more unknots which are intertwined such that they are not separable.

11.5.1 Hopf Link

$$x = \pm a + \cos u$$
$$y = \sin u$$
$$z = \pm b \sin u,$$

where each ring is of unit radius, offset by a distance c from the origin, and b is the height to which each individual unknot is slanted in order that they interlock. The plus or minus sign is used for the right and left unknot, respectively.

$a=0.5$, $b=0.2$; $0<u<2\pi$; viewpoint=(0,0,10)

11.4.8a

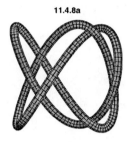

11.4.8b

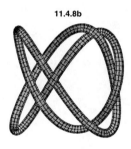

11.5.1

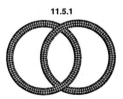

11.5.2 Borromean Rings Link

A third unknot is added to the Hopf link above to form the Borremean rings. The three unknots are given parametrically by:

left ring	right ring	lower ring
$x = -a + \cos u$	$x = a + \cos u$	$x = \cos u$
$y = \sin u$	$y = \sin u$	$y = \sin u - \sin(\pi/3)$
$z = b \cos(3u)$	$z = b \cos(3u)$	$z = b \cos(3u)$

$a = 0.5$, $b = 0.2$; $0 < u < 2\pi$; viewpoint $= (0,0,10)$

11.5.3 Whitehead Link

The Whitehead link is formed of two unknots, one of which has radius b and is given a half-twist to height d while the other simple unknot of radius c links them. These two unknots are given parametrically by:

twisted unknot	simple unknot
$x = b \sin u$	$x = a \cos u$
$y = b \sin(2u)/2$	$y = a \sin u$
$z = -d \cos u$	$z = \sin(2u - 3\pi/4)$

$a = 0.5$, $b = 1.0$, $d = 0.1$; $0 < u < 2\pi$; viewpoint $= (0,0,10)$

References

1. J.D. Foley and A. VanDam. 1983. *Fundamentals of Interactive Computer Graphics*, Reading, MA: Addison-Wesley.
2. R.B. Thompson. 1998. "Designing a baseball cover," *College Mathematics Journal*, 29, pp. 48–61.

11.5.2

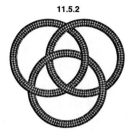

11.5.3

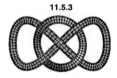

12

Algebraic Surfaces

The following forms are plotted in the perspective projection described at the beginning of Chapter 11. There are numerous, different ways to represent 3-D surfaces. The method chosen here is a shaded-relief type of illustration, with a grid superimposed to show lines of constant x and y on the surface. Artificial light sources are used to enhance the features of each surface, and so the shading is not a uniform gray. The viewpoint of the observer is given as (x, y, z) coordinates relative to the origin of the projection plane that is coincident with the origin of the surface and normal to the line connecting the origin with the observer.

The surfaces are shown in their true aspect ratios, and a bounding box is placed about the surface in each case. This box extends from -1 to $+1$ on both the x and y axes. The z range is also -1 to $+1$, with a few exceptions as indicated. The surfaces sometimes intersect the top or bottom of the box if they have become unbounded; at these intersections, the surface is not always represented accurately by the graphical method used here.

12.1 Functions with $ax + by$

12.1.1 $z = ax + by$ $ax + by - z = 0$

Plane

   a. $a = 0.5$, $b = 0.5$; viewpoint $= (5, -6, 4)$

   b. $a = 0.1$, $b = 0.3$; viewpoint $= (5, -6, 4)$

12.1.2 $z = 1/(ax + by)$ $axz + byz - 1 = 0$

   a. $a = 5.0$, $b = 5.0$; viewpoint $= (6, -4, 3)$

   b. $a = 2.0$, $b = 4.0$; viewpoint $= (6, -1, 3)$

12.1.1a

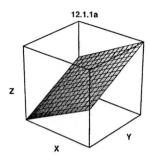

12.1.1b

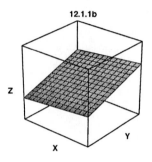

12.1.2a

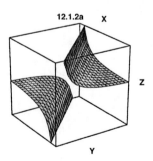

12.1.2b

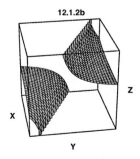

12.2   Functions with $x^2/a^2 \pm y^2/b^2$

12.2.1 $z = c(x^2/a^2 + y^2/b^2)$ $x^2/a^2 + y^2/b^2 - z/c = 0$

Elliptic paraboloid

  a. $a = 0.5$, $b = 1.0$, $c = -1.0$; viewpoint$= (5, -6, 4)$

  b. $a = 1.0$, $b = 1.0$, $c = -2.0$; viewpoint$= (5, -6, 4)$

12.2.2 $z = c(x^2/a^2 - y^2/b^2)$ $x^2/a^2 - y^2/b^2 - z/c = 0$

Hyperbolic paraboloid (also called *saddle*)

  a. $a = 0.5$, $b = 0.5$, $c = 1.0$; viewpoint$= (4, -6, 4)$

  b. $a = 1.0$, $b = 0.5$, $c = 1.0$; viewpoint$= (4, -6, 4)$

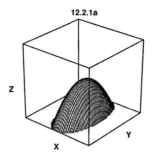

12.2.1a

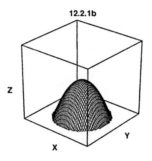

12.2.1b

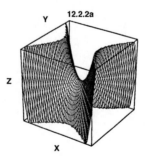

12.2.2a

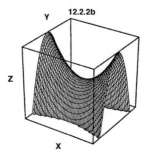

12.2.2b

12.2.3 $1 = x^2/a^2 + y^2/b^2$ $x^2/a^2 + y^2/b^2 - 1 = 0$

Elliptic cylinder
$a = 0.5$, $b = 1.0$; viewpoint $= (4, -5, 2)$

12.2.4 $1 = x^2/a^2 - y^2/b^2$ $x^2/a^2 - y^2/b^2 - 1 = 0$

Hyperbolic cylinder
$a = 0.25$, $b = 0.25$; viewpoint $= (4, -6, 3)$

12.3   Functions with $(x^2/a^2 + y^2/b^2 \pm c^2)^{1/2}$

12.3.1 $z = (1 - x^2 - y^2)^{1/2}$ $x^2 + y^2 + z^2 - 1 = 0$

Sphere
viewpoint $= (4, -5, 2)$

12.3.2 $z = c(1 - x^2/a^2 - y^2/b^2)^{1/2}$ $x^2/a^2 + y^2/b^2 + z^2/c^2 - 1 = 0$

Ellipsoid
Special cases: $a = b > c$ gives *oblate spheroid*; $a = b < c$ gives *prolate spheroid*

   a. $a = 1.0$, $b = 1.0$, $c = 0.5$; viewpoint $= (4, -5, 2)$

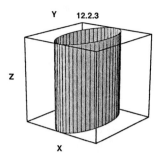

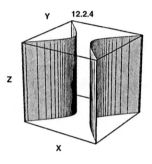

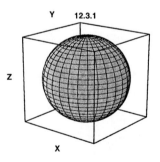

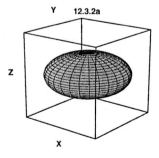

b. $a=0.5$, $b=0.5$, $c=1.0$; viewpoint$=(4,-5,2)$

12.3.3 $z = c(x^2 + y^2)^{1/2}$ $x^2 + y^2 - z^2 = 0$

Cone

a. $c=0.5$; viewpoint$=(4,-5,2)$

b. $c=2.0$; viewpoint$=(4,-5,2)$

12.3.4 $z = c(x^2/a^2 + y^2/b^2)^{1/2}$ $x^2/a^2 + y^2/b^2 - z^2/c^2 = 0$

Elliptic cone (circular cone if $a=b$)

a. $a=0.5$, $b=1.0$, $c=1.0$; viewpoint$=(4,-5,2)$

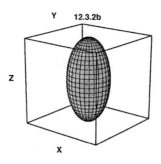

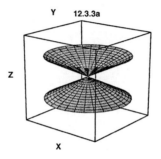

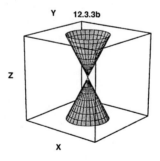

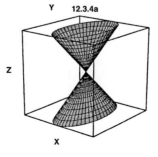

b. $a=0.5$, $b=1.0$, $c=2.0$; viewpoint$=(4,-5,2)$

12.3.5 $z = c(x^2/a^2 + y^2/b^2 - 1)^{1/2}$ $x^2/a^2 + y^2/b^2 - z^2/c^2 - 1 = 0$

Hyperboloid of one sheet

a. $a=0.2$, $b=0.2$, $c=0.2$; viewpoint$=(4,-5,2)$

b. $a=0.2$, $b=0.2$, $c=0.4$; viewpoint$=(4,-5,2)$

12.3.6 $z = c(x^2/a^2 + y^2/b^2 + 1)^{1/2}$ $x^2/a^2 + y^2/b^2 - z^2/c^2 + 1 = 0$

Hyperboloid of two sheets

a. $a=0.2$, $b=0.2$, $c=0.1$; viewpoint$=(4,-5,2)$

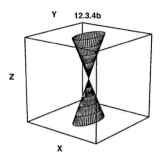

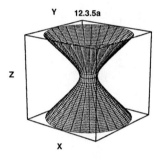

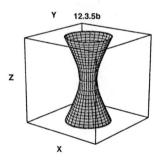

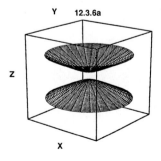

b. $a=0.2$, $b=0.2$, $c=0.2$; viewpoint$=(4,-5,2)$

12.4   Functions with $x^3/a^3 \pm y^3/b^3$

12.4.1 $z = c(x^3/a^3 + y^3/b^3)$ $x^3/a^3 + y^3/b^3 - z/c = 0$

a. $a=1.0$, $b=1.0$, $c=0.5$; viewpoint$=(1,-5,2)$

b. $a=1.0$, $b=1.0$, $c=2.0$; viewpoint$=(1,-5,2)$

12.4.2 $z = c(x^3/a^3 - y^3/b^3)$ $x^3/a^3 - y^3/b^3 - z/c = 0$

a. $a=1.0$, $b=1.0$, $c=0.5$; viewpoint$=(1,-5,2)$

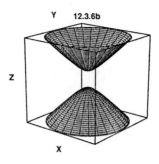

12.3.6b

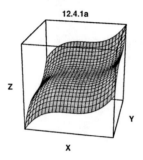

12.4.1a

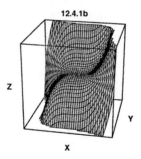

12.4.1b

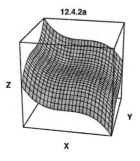

12.4.2a

b. $a=1.0$, $b=1.0$, $c=2.0$; viewpoint$=(1,-5,2)$

12.5   Functions with $x^4/a^4 \pm y^4/b^4$

12.5.1 $z = c(x^4/a^4 - y^4/b^4)$ $x^4/a^4 - y^4/b^4 - z/c = 0$

a. $a=1.0$, $b=1.0$, $c=0.5$; viewpoint$=(-4,3,2)$

b. $a=1.0$, $b=1.0$, $c=2.0$; viewpoint$=(2,-4,1)$

12.5.2 $z = c(x^4/a^4 - y^4/b^4)$ $x^4/a^4 - y^4/b^4 - z/c = 0$

a. $a=1.0$, $b=1.0$, $c=0.5$; viewpoint$=(-4,3,2)$

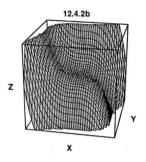

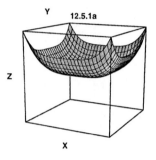

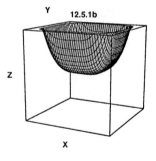

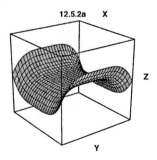

b. $a=1.0$, $b=1.0$, $c=2.0$; viewpoint$=(-4,3,2)$

12.6 Miscellaneous Functions

$$z = (a^2 - [(x^2+y^2)^{1/2} - b]^2)^{1/2}$$

12.6.1

$$x^4 + y^4 + 2x^2y^2 + 2(a^2-b^2)(x^2+y^2) + z^4 - 2z^2(a^2+b^2) + (a^2+b^2)^2 = 0$$

Torus ($a=b$ gives *spindle torus*)

a. $a=0.8$, $b=0.2$; viewpoint$=(4,-5,2)$

b. $a=0.5$, $b=0.5$; viewpoint$=(4,-5,2)$

12.6.2 $z = c(x^3 - 3xy^2)$ $x^3 - 3xy^2 - z/c = 0$

Monkey saddle
$c=1.0$; viewpoint$=(5,3,3)$

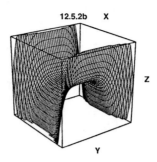

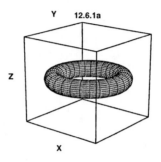

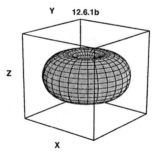

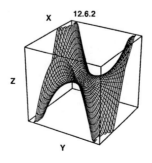

12.6.3 $z = cxy^2$ $cxy^2 - z = 0$

$c = 1.0$; viewpoint $= (5, -3, 3)$

12.6.4 $z = cx^2y^2$ $cx^2y^2 - z = 0$

Crossed trough
$c = 4.0$; viewpoint $= (2, -5, 3)$

12.6.5 $z = (ax^2 + by^2)/(x^2 + y^2)$ $ax^2 + by^2 - zx^2 - zy^2 = 0$

Conoid of Plucker or *cylindroid*
$a = -0.5$, $b = 0.8$; viewpoint $= (5, -3, 2)$

12.6.6 $z = x(ax^2 + by^2)$ $ax^3 + bxy^2 - z = 0$

   a. $a = 1.5$, $b = 1.0$; viewpoint $= (-3, -4, 2)$

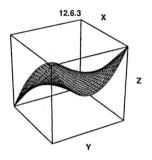

12.6.3

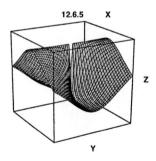

12.6.4

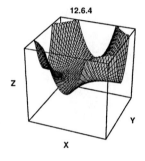

12.6.5

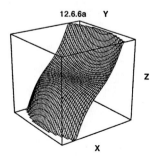

12.6.6a

b. $a=1.5$, $b=3.0$; viewpoint$=(-3,-4,2)$

12.6.7 $z = \pm\left[a^n - (|x|^n + |y|^n)\right]^{1/n}$ $z^n \pm \left[a^n - (|x|^n + |y|^n)\right] = 0$

Hypersphere ($n=2$ gives ordinary sphere)

a. $a=1.0$, $n=3$; viewpoint$=(-3,-4,2)$

b. $a=1.0$, $n=5$; viewpoint$=(-3,-4,2)$

12.6.8 $z = cx^2/y^2$ $cx^2 - zy^2 = 0$

Whitney's umbrella
$c=1.0$; viewpoint$=(-2,-4,3)$

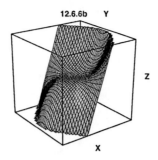

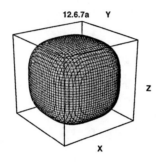

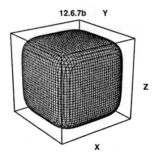

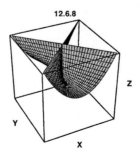

12.6.9 $z = cxy/(x^2 + y^2)$ $cxy - z(x^2 + y^2) = 0$

c=1.0; viewpoint=(4,−2,3)

12.7 Miscellaneous Functions Expressed Parametrically

Eight Surface

12.7.1   Implicit equation : $a^2(x^2 + y^2) - z^2 - z^4 = 0$

Defined parametrically by:

$$x = \cos u \cos v \sin v/a$$
$$y = \sin u \cos v \sin v/a$$
$$z = \sin v.$$

a=0.5; 0<u<2π; −π/2<v<π/2; viewpoint=(−3,−4,2)

Goursat's Surface

12.7.2   Implicit equation : $x^4 + y^4 + z^4 + a(x^2 + y^2 + z^2)^2 - c = 0.$

Goursat's surface is composed of two surfaces. Let c=1+a; then the two surfaces are:

$$x^4 + y^4 + z^4 = 1 \quad \textit{(hypersphere; see Section 12.6.7)}$$
$$a(x^2 + y^2 + z^2)^2 = a.$$

Parametrically, these two equations are:

$$x = [\cos \theta \sin \varphi]^{1/2} \quad x = \cos \theta \sin \varphi$$
$$y = [\sin \theta \sin \varphi]^{1/2} \quad y = \sin \theta \sin \varphi$$
$$z = [\cos \varphi]^{1/2} \quad z = \cos \varphi$$

a. a=0.3, c=0.8; 0<θ<π and 0<φ<π; viewpoint=(−3,−4,2)

b. a=−0.3, c=0.2; 0<θ<2π and 0<φ<π; viewpoint=(−3,−4,2)

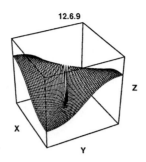

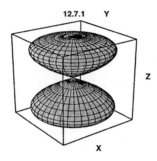

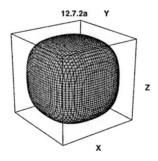

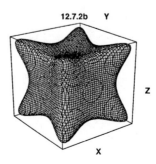

Moebius Strip
Defined parametrically by:

$$x = \cos u[1 + v \cos(u/2)]$$

12.7.3 $$y = \sin u[1 + v \cos(u/2)]$$

$$z = v \sin(u/2).$$

$a = 0.2$, $0 < u < 2\pi$; $-a < v < a$; axes limits $= \pm(1+a)$; viewpoint $= (-2, -4, 3)$

Crosscap
Defined parametrically by:

$$x = \sin u \, \sin(2v)/2$$

12.7.4 $$y = \sin(2u)\cos^2 v$$

$$z = \cos(2u)\cos^2 v$$

$0 < u < \pi$; $-\pi/2 < v < \pi/2$; viewpoint $= (-3, -4, 2)$

Bohemian Dome
Defined parametrically by:

$$x = a \cos u$$

12.7.5 $$y = a \sin u + b \cos v$$

$$z = c \sin v$$

Note: figure shows only the bottom half of the surface that is symmetric about the z plane.
$a = 0.75$, $b = 0.25$, $c = 0.25$; $0 < u < 2\pi$; $0 < v < 2\pi$; $-1 < z < 0$; viewpoint $= (-4, -2, 3)$

Conical Wedge
$a = 1$, $b = \sqrt{2}$ gives Conocuneus of Wallis.

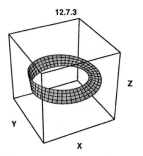

12.7.3

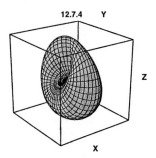

12.7.4 Y

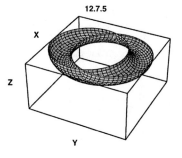

12.7.5

Defined parametrically by

12.7.6
$$x = v \cos u$$
$$y = v \sin u$$
$$z = c[a^2 - b^2\cos^2 u]^{1/2}.$$

    a. $a=1.0$, $b=1.0$, $c=1.0$; $0<u<\pi$; $-1<v<1$; $0<z<1$; viewpoint$=(-3,-5,4)$

    b. $a=1.0$, $b=\sqrt{2}$, $c=1.0$; $\pi/4<u<3\pi/4$; $-1<v<1$; $0<z<1$;
       viewpoint$=(-3,-5,4)$

Astroidial Surface

12.7.7   Implicit equation : $x^{2/3} + y^{2/3} + z^{2/3} = a^2$

Defined parametrically by:

$$x = [a \cos u \cos v]^3$$
$$y = [a \sin u \cos v]^3$$
$$z = [a \sin v]^3$$

$0<u<2\pi$; $-\pi/2<v<\pi/2$
$a=1.0$; $0<u<2\pi$; $-\pi/2<v<\pi/2$; viewpoint$=(-4,-3,2)$

12.7.6a

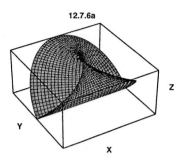

12.7.6b

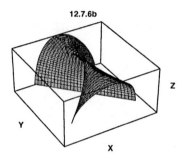

12.7.7

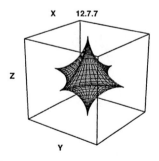

Corkscrew Surface
Also called *twisted sphere*.
Defined parametrically by:

$$x = a \cos u \cos v$$

12.7.8 $$y = a \sin u \cos v$$

$$z = a \sin v + bu.$$

$a=1.0$, $b=2/\pi$; $0<u<2\pi$; $-\pi<v<\pi$; $-a<z<a+2\pi b$; viewpoint$=(-2,3,2)$

Dini's Surface
Defined parametrically by:

$$x = a \cos u \sin v$$

12.7.9 $$y = a \sin u \sin v$$

$$z = a\{\cos v + \ln[\tan(v/2)]\} + bu.$$

$a=1.0$, $b=1/(2\pi)$; $2\pi<u<6\pi$; $0<v<\pi/2$; $0<z<3$; viewpoint$=(-2,-5,3)$

Cyclide of Dupin
Defined parametrically by:

$$x = \{c[b-c \cos u] + a \cos u[a-b \cos v]\}/[a-c \cos u \cos v]$$

12.7.10 $$y = \{(a^2 - c^2)^{1/2}[a-b \cos v]\sin u\}/[a-c \cos u \cos v]$$

$$z = \{(a^2 - c^2)^{1/2}[b-c \cos u]\sin v\}/[a-c \cos u \cos v].$$

Note: For $b=c$, the figure has an infinitesimal cross-section at $u=0$; for $c > b$, the figure has two such cross-sections.

a. $a=0.6$, $b=0.2$, $c=0.1$; $0<u<2\pi$; $0<v<2\pi$; viewpoint$=(2,-5,3)$

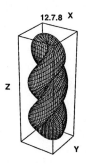

12.7.8

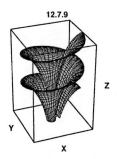

12.7.9

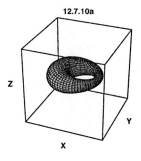

12.7.10a

b. $a=0.6$, $b=0.2$, $c=0.2$; $0<u<2\pi$; $0<v<2\pi$; viewpoint$=(2,-5,3)$

c. $a=0.6$, $b=0.2$, $c=0.3$; $0<u<2\pi$; $0<v<2\pi$; viewpoint$=(2,-5,3)$

Sine Surface
Defined parametrically by:

$$x = a \sin u$$
12.7.11 $\quad\quad y = a \sin v$
$$z = a \sin(u + v).$$

$a=0.75$; $0<u<2\pi$; $-\pi<v<\pi$; viewpoint$=(-3,-4,3)$

Cosine Surface
Defined parametrically by:

$$x = a \cos u$$
12.7.12 $\quad\quad y = a \cos v$
$$z = a \cos(u + v).$$

$a=0.75$; $0<u<2\pi$; $-\pi<v<\pi$; viewpoint$=(-2,-4,3)$

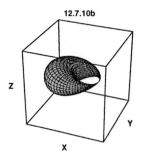

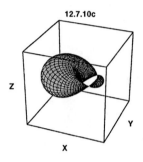

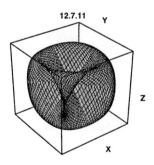

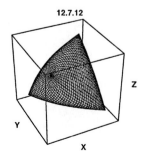

Catenoid
Defined parametrically by:

$$x = au \cos v$$

12.7.13 $$y = au \sin v$$

$$z = \pm a[\operatorname{arccosh}(u)].$$

$a = 0.2$; $0 < u < 5$; $0 < v < 2\pi$; viewpoint $= (4, -5, 1)$

Right Helicoid
Defined parametrically by:

$$x = u \cos v$$

12.7.14 $$y = u \sin v$$

$$z = cv$$

a. $c = 1/(2\pi)$; $0.0 < u < 1.0$; $-2\pi < v < 2\pi$; viewpoint $= (4, -5, 1)$

b. $c = 1/(4\pi)$; $0.0 < u < 1.0$; $-4\pi < v < 4\pi$; viewpoint $= (4, -5, 1)$

c. $c = 1/(2\pi)$; $0.2 < u < 0.8$; $-2\pi < v < 2\pi$; viewpoint $= (4, -5, 1)$

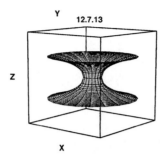

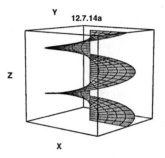

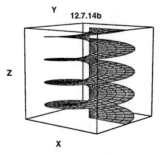

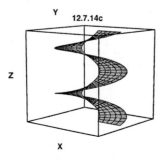

Fresnel's Elasticity Surface

12.7.15 $r = (a^2x^2 + b^2y^2 + c^2z^2)^{1/2}$

Defined parametrically by:

$$x = r \sin u \cos v/a$$
$$y = r \sin u \sin v/b$$
$$z = r \cos u/c.$$

$a = 0.5$, $b = 1.0$, $c = 0.5$; $0 < u < \pi$; $0 < v < 2\pi$; viewpoint $= (4, -3, 3)$

Cornucopia
Defined parametrically by:

$$x = c[e^{bv}\cos v + e^{av}\cos u \cos v]$$
12.7.16 $y = c[e^{bv}\sin v + e^{av}\cos u \sin v]$
$$z = c[e^{av}\sin u].$$

$a = 0.25$, $b = 0.40$, $c = 0.45$; $0 < u < 2\pi$; $-5 < v < 1$; viewpoint $= (4, -3, 3)$

Cylindrical Spiral
Defined parametrically by:

$$x = a \cos(nv)[1 + \cos u] + c \cos(nv)$$
12.7.17 $y = a \sin(nv)[1 + \cos u] + c \sin(nv)$
$$z = bv/(2\pi) + a \sin u.$$

a. $a = 0.1$, $b = 1.0$, $c = 0.7$, $n = 3$; $0 < u < 2\pi$; $-\pi < v < \pi$; viewpoint $= (2, -4, 2)$

b. $a = 0.3$, $b = 1.0$, $c = 0.4$, $n = 3$; $0 < u < 2\pi$; $-\pi < v < \pi$; viewpoint $= (2, -4, 2)$

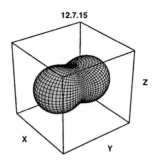

12.7.15

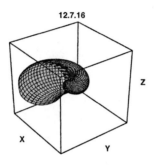

12.7.16

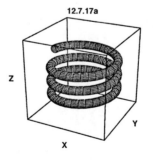

12.7.17a

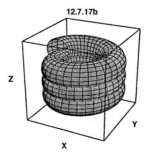

12.7.17b

Conical Spiral
Defined parametrically by:

$$x = a[1 - v/(2\pi)]\cos(nv)[1 + \cos u] + c\cos(nv)$$

12.7.18
$$y = a[1 - v/(2\pi)]\sin(nv)[1 + \cos u] + c\sin(nv)$$

$$z = bv/(2\pi) + a[1 - v/(2\pi)]\sin u.$$

a. $a = 0.2$, $b = 1.0$, $c = 0.1$, $n = 2$; $0 < u < 2\pi$; $-\pi < v < 2\pi$; viewpoint $= (2, -4, 2)$

b. $a = 0.3$, $b = 0.6$, $c = 0.1$, $n = 2$; $0 < u < 2\pi$; $-\pi < v < 2\pi$; viewpoint $= (2, -4, 2)$

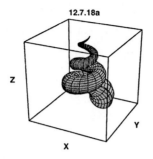

12.7.18a

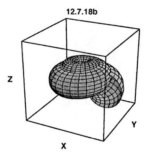

12.7.18b

13

Transcendental Surfaces

The perspective projection described at the beginning of Chapter 11 is used for the figures in this chapter. The surface representation is as described at the beginning of Chapter 12. The enclosing box, shown with most figures of this chapter, has limits of -1 to $+1$ for all three axes, unless noted otherwise. The true aspect ratio of all figures is preserved. Wherever the surfaces become unbounded and intersect the top or bottom of the enclosing box, they are not always accurately represented by the graphical method used here.

13.1   Trigonometric Functions

13.1.1 $z = c \sin[2\pi a(x^2 + y^2)^{1/2}]$

  $a = 2.0$, $c = 0.25$; viewpoint $= (5, -4, 4)$

13.1.2 $z = c \cos[2\pi a(x^2 + y^2)^{1/2}]$

  $a = 2.0$, $c = 0.25$; viewpoint $= (5, -4, 4)$

13.1.3 $z = c \sin(2\pi a x y)$

  $a = 3.0$, $c = 0.25$; viewpoint $= (5, -4, 4)$

13.1.4 $z = c \cos(2\pi a x y)$

  $a = 3.0$, $c = 0.25$; viewpoint $= (5, -4, 4)$

13.1.1

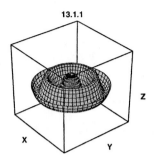

13.1.2

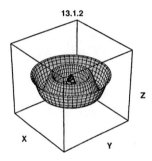

13.1.3

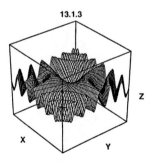

13.1.4

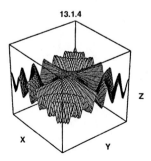

13.1.5 $z = c \, \sin(2\pi ax)\sin(2\pi by)$

   $a=2.0, \, b=1.0, \, c=0.5$; viewpoint$=(5,-4,4)$

13.1.6 $z = c \, \cos(2\pi ax)\cos(2\pi by)$

   $a=2.0, \, b=1.0, \, c=0.5$; viewpoint$=(5,-4,4)$

13.2 Logarithmic Functions

13.2.1 $z = c \, \ln(a|x| + b|y|)$

   $a=1.0, \, b=2.0, \, c=0.5$; viewpoint$=(4,-3,4)$

13.2.2 $z = c \, \ln(ax^2 + by^2)$

   $a=1.0, \, b=1.0, \, c=0.7; \; -2<z<0$; viewpoint$=(-2,-4,2)$

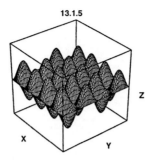

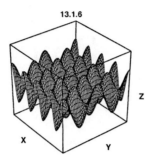

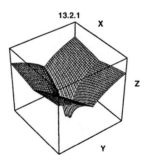

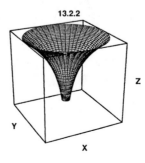

13.2.3 $z = c \ln(|xy|)$

$c = 0.2$; viewpoint $= (4, -3, 4)$

13.3 Exponential Functions

13.3.1 $z = c e^{ax+by}$

$a = 2.0$, $b = 2.0$, $c = 0.1$; viewpoint $= (-3, -5, 3)$

13.3.2 $z = c e^{ax^2 + by^2}$

a. $a = 1.0$, $b = 0.5$, $c = 0.25$; viewpoint $= (5, -3, 3)$

b. $a = 1.0$, $b = -0.5$, $c = 0.25$; viewpoint $= (5, -3, 3)$

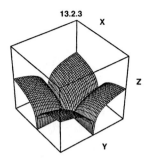

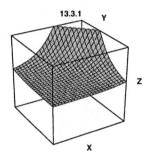

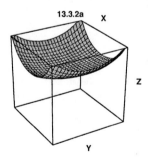

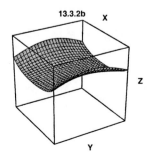

c. $a=-2.0$, $b=-1.0$, $c=1.00$; viewpoint$=(5,-3,3)$

13.3.3 $z = ce^{axy}$

$a=2.0$, $c=0.35$; viewpoint$=(5,-3,2)$

13.4 Trigonometric and Exponential Functions Combined

13.4.1 $z = c\cos[2\pi a(x^2+y^2)^{1/2}]e^{-b(x^2+y^2)^{1/2}}$

$a=3.0$, $b=2.0$, $c=0.5$; viewpoint$=(3,2,3)$

13.4.2 $z = c\sin[2\pi a(x^2+y^2)^{1/2}]e^{-b(x^2+y^2)^{1/2}}$

$a=3.0$, $b=2.0$, $c=0.5$; viewpoint$=(3,2,3)$

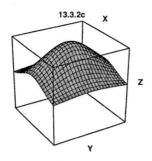

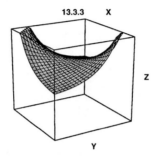

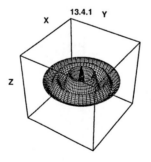

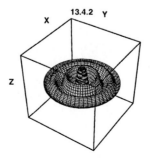

13.4.3 $z = c \cos(2\pi bx)e^{ay}$

$a=1.0$, $b=1.0$, $c=0.4$; viewpoint$=(3,-4,3)$

13.4.4 $z = c \sin(2\pi bx)e^{ay}$

$a=1.0$, $b=1.0$, $c=0.4$; viewpoint$=(3,-4,3)$

13.5 Surface Spherical Harmonics

Surface spherical harmonics are deformations of the sphere which are periodic in the polar angle ϕ or the longitudinal angle θ. The arbitrary sphere is assigned a radius of unity here. For these harmonics, one-half of the surfaces are cut away along the x–z plane in order to more completely illustrate the shape of these surfaces. The three orthogonal axes, all from -1 to $+1$, are added to also clarify the figure.

13.5.1 $r = 1 + cP_n^0(\cos \phi)$

Zonal harmonics: P_n^0 is the Legendre polynomial (see Section 4.1)

   a. $n=1$, $c=1.0$; viewpoint$=(2,-3,2)$

   b. $n=2$, $c=1.0$; viewpoint$=(2,-3,2)$

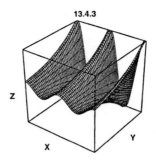

13.4.3

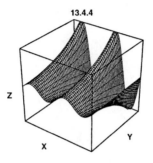

13.4.4

13.5.1a

13.5.1b

c. $n=3$, $c=1.0$; viewpoint$=(2,-3,2)$

13.5.2 $r = 1 + cP_n^n(\cos \phi)\cos(n\theta)$

Sectoral harmonics: P_n^n is the associated Legendre function of the first kind (see Section 4.1)

a. $n=1$, $c=1.0$; viewpoint$=(3,-2,2)$

b. $n=2$, $c=1/3$; viewpoint$=(3,-2,2)$

c. $n=3$, $c=1/15$; viewpoint$=(3,-2,2)$

13.5.1c

13.5.2a

13.5.2b

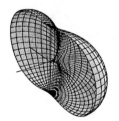

13.5.2c

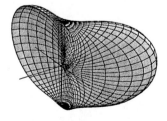

13.5.3 $r = 1 + cP_n^m(\cos \phi)\cos(m\theta)$

Tesseral harmonics: P_n^m is the associated Legendre function of the first kind (see Section 4.1)

  a. $n=2$, $m=1$, $c=2/3$; viewpoint$=(2,-3,2)$

  b. $n=2$, $m=2$, $c=1/3$; viewpoint$=(2,-3,2)$

  c. $n=3$, $m=1$, $c=2/5$; viewpoint$=(2,-3,2)$

  d. $n=3$, $m=2$, $c=1/6$; viewpoint$=(2,-3,2)$

13.5.3a

13.5.3b

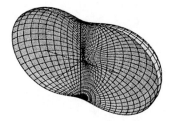

13.5.3c

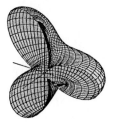

13.5.3d

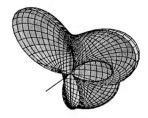

14

Complex Variable Surfaces

The functions of this chapter are given by $w = f(z)$ where z is the complex number $x + iy$. To illustrate the functions, both abs(w) and arg(w) are plotted, where

$$\text{abs}(w) = (u^2 + v^2)^{1/2}$$

$$\text{arg}(w) = \arctan(v/u)$$

$$u = \text{real}(z)$$

$$v = \text{imaginary}(z).$$

The perspective projection described at the beginning of Chapter 11 is used for the figures in this chapter. The surface representation is as described at the beginning of Chapter 12.

The enclosing box shown with all abs(w) figures of this chapter has limits of -1 to $+1$ for x and y axes, but limits of 0 to $+1$ for the $z = \text{abs}(w)$ axis. The true aspect ratio of all abs(w) figures is preserved. For arg(w) plots, the bounding box also has limits of -1 to $+1$ for x and y axes but limits of $-\pi$ to $+\pi$ for the $z = \text{arg}(w)$ axis. Many of the arg(w) plots exhibit jumps of 2π in the complex plane, along branch cuts of the particular function.

14.1 Algebraic Functions

14.1.1 $\qquad w = cz$

   a. $\mathrm{abs}(w)$; $c = 1/\sqrt{2}$; viewpoint $= (-1, -4, 2)$

   b. $\arg(w)$; $c = 1/\sqrt{2}$; viewpoint $= (-1, -4, 2)$

14.1.2 $\qquad w = cz^2$

   a. $\mathrm{abs}(w)$; $c = 0.5$; viewpoint $= (-2, -4, 2)$

   b. $\arg(w)$; $c = 0.5$; viewpoint $= (-2, -4, 2)$

14.1.1a

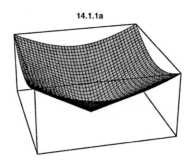

14.1.1b

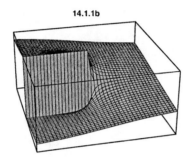

14.1.2a

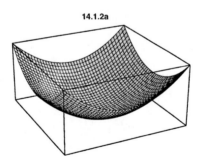

14.1.2b

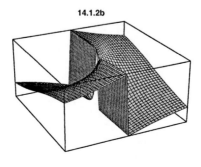

14.1.3 $w = c/z$

  a. $\text{abs}(w)$; $c = 0.1$; viewpoint $= (5,3,2)$

  b. $\arg(w)$; $c = 0.1$; viewpoint $= (5,3,2)$

14.1.4 $w = c/z^2$

  a. $\text{abs}(w)$; $c = 0.01$; viewpoint $= (4,3,3)$

  b. $\arg(w)$; $c = 0.01$; viewpoint $= (4,3,3)$

14.1.3a

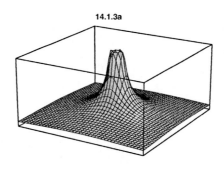

14.1.3b

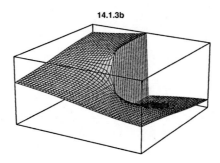

14.1.4a

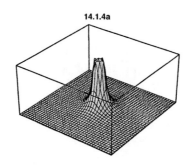

14.1.4b

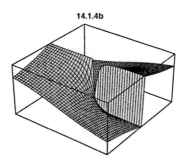

14.1.5 $w = az + b$

   a. abs(w); $a = 0.5$, $b = 0.25$; viewpoint $= (3, -4, 4)$

   b. arg(w); $a = 0.5$, $b = 0.25$; viewpoint $= (3, -4, 4)$

14.1.6 $w = c/(z - a)$

   a. abs(w); $a = 0.5$, $c = 0.1$; viewpoint $= (-1, 4, 3)$

   b. arg(w); $a = 0.5$, $c = 0.1$; viewpoint $= (-1, 4, 3)$

14.1.5a

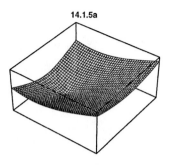

14.1.5b

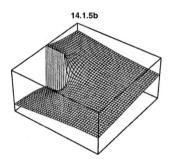

14.1.6a

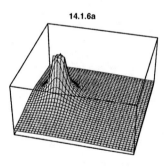

14.1.6b

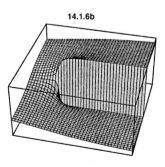

14.1.7 $w = c/(z-a)^2$

a. abs(w); $a=0.5$, $c=0.01$; viewpoint$=(-1,4,3)$

b. arg(w); $a=0.5$, $c=0.01$; viewpoint$=(-1,4,3)$

14.1.8 $w = cz/(z-a)$

a. abs(w); $a=0.5$, $c=0.25$; viewpoint$=(-1,4,3)$

b. arg(w); $a=0.5$, $c=0.25$; viewpoint$=(-1,4,3)$

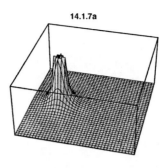

14.1.7a

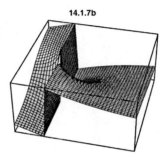

14.1.7b

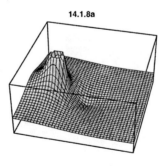

14.1.8a

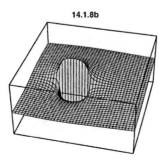

14.1.8b

14.1.9 $w = c/(z^2 - a^2)$

   a. abs(w); $a=0.5$, $c=0.05$; viewpoint$=(-1,4,3)$

   b. arg(w); $a=0.5$, $c=0.05$; viewpoint$=(-1,4,3)$

14.1.10 $w = c(z + 1/z)$

   a. abs(w); $c=0.1$; viewpoint$=(4,2,3)$

   b. arg(w); $c=0.1$; viewpoint$=(4,2,3)$

14.1.9a

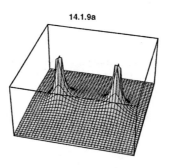

14.1.9b

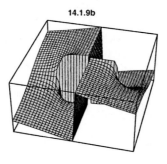

14.1.10a

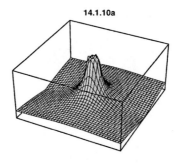

14.1.10b

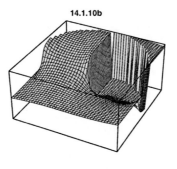

14.1.11 $w = (az + b)/(cz + d)$

  a. abs(w); $a = -0.2$, $b = 0.1$, $c = 1.2$, $d = 0.6$; viewpoint $= (4,2,3)$

  b. arg(w); $a = -0.2$, $b = 0.1$, $c = 1.2$, $d = 0.6$; viewpoint $= (4,2,3)$

14.2 Transcendental Functions

14.2.1 $w = ce^z$

  a. abs(w); $c = 0.3$; viewpoint $= (-4,-2,3)$

  b. arg(w); $c = 0.3$; viewpoint $= (-4,-2,3)$

14.1.11a

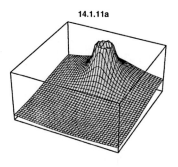

14.1.11b

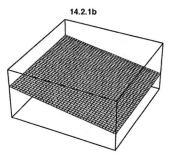

14.2.1a

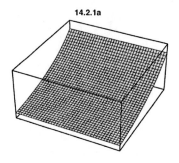

14.2.1b

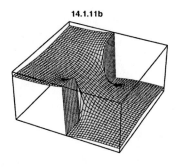

14.2.2 $w = c \ln(z)$

    a. abs(w); $c=0.25$; viewpoint$=(4,-2,3)$

    b. arg(w); $c=0.25$; viewpoint$=(4,-2,3)$

14.2.3 $w = c \sin(\pi z)$

    a. abs(w); $c=0.1$; viewpoint$=(4,-2,3)$

    b. arg(w); $c=0.1$; viewpoint$=(4,-2,3)$

14.2.2a

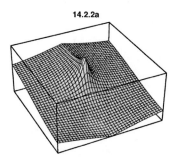

14.2.2b

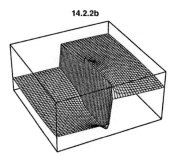

14.2.3a

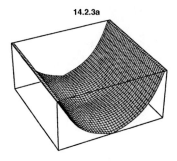

14.2.3b

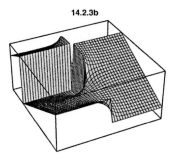

14.2.4 $w = c \cos(\pi z)$

     a. abs(w); $c = 0.1$; viewpoint $= (4, -2, 3)$

     b. arg(w); $c = 0.1$; viewpoint $= (4, -2, 3)$

14.2.5 $w = c \tan(\pi z)$

     a. abs(w); $c = 0.1$; viewpoint $= (4, -2, 3)$

     b. arg(w); $c = 0.1$; viewpoint $= (4, -2, 3)$

14.2.4a

14.2.4b

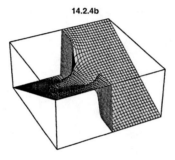

14.2.5a

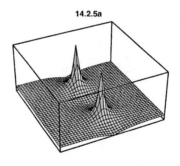

14.2.5b

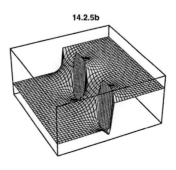

14.2.6 $w = c \sinh(z)$

    a. abs(w); $c = 0.5$; viewpoint $= (-2, -4, 2)$

    b. arg(w); $c = 0.5$; viewpoint $= (-2, -4, 2)$

14.2.7 $w = c \cosh(z)$

    a. abs(w); $c = 0.5$; viewpoint $= (-2, -4, 2)$

    b. arg(w); $c = 0.5$; viewpoint $= (-2, -4, 2)$

14.2.6a

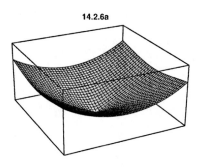

14.2.6b

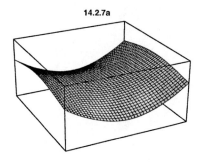

14.2.7a

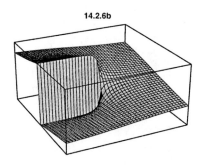

14.2.7b

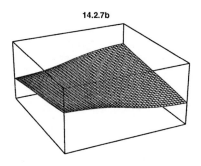

14.2.8 $w = c \tanh(5z)$

  a. abs(w); $c=0.5$; viewpoint$=(2,-4,3)$

  b. arg(w); $c=0.5$; viewpoint$=(2,-4,3)$

14.2.8a

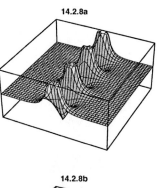

14.2.8b

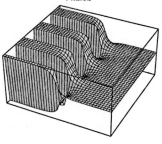

15

Minimal Surfaces

The field of minimal surfaces is an active and growing one.[1] A minimal surface is defined as having zero mean curvature.[2] The plane is the first and simplest of the family of minimal surfaces, with the catenoid and helicoid being the earliest non-trivial examples found. This chapter presents only some of the more commonly known minimal surfaces rather than attempting to be exhaustive.

The projection used here is the perspective one described at the beginning of Chapter 11. The surface representation is as described at the beginning of Chapter 12. The enclosing box shown with the figures of this chapter has limits of -1 to $+1$ for all three axes, unless noted otherwise.

15.1 Elementary Minimal Surfaces

15.1.1 Catenoid (Also Seen as 12.7.13)

Defined parametrically by:

$$x = u \cos v,$$
$$y = u \sin v,$$
$$z = \pm \text{arccosh}(u).$$

$1 < u < 5, 0 < v < 2\pi$; viewpoint $= (4, -5, 1)$

15.1.2 Right Helicoid (Also Seen as 12.7.14)

Defined parametrically by:

$$x = u \cos v,$$
$$y = u \sin v,$$
$$z = cv.$$

$c = 1/(2\pi)$; $-0.8 < u < 0.8$; $-2\pi < v < 2\pi$; viewpoint $= (4, -5, 1)$

15.2 Complex Minimal Surfaces

15.2.1 Enneper's Surface

Defined parametrically by

$$x = r \cos(\varphi) - \frac{r^3}{3} \cos(3\varphi)$$

$$y = -\frac{r}{3}[3 \sin(\varphi) + r^2 \sin(3\varphi)]$$

$$z = r^2 \cos(2\varphi).$$

$0 < r < 2$, $-\pi < \varphi < \pi$; $-4 < x < 4$, $-4 < y < 4$, $-4 < z < 4$; viewpoint $= (4, -4, 2)$

15.2.2 Costa's Surface

See http://mathworld.wolfram.com/CostaMinimalSurface.html for a description of the construction of Costa's surface.

$0 < u < 1, 0 < v < 1$; $-4 < x < 4$, $-4 < y < 4$, $-3 < z < 3$; viewpoint $= (4, -3, 3)$

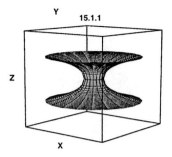

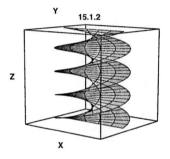

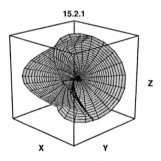

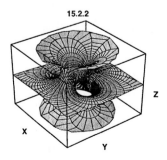

15.2.3 Hennenberg's Surface

Defined parametrically by:

$$x = 2 \sinh u \cos v - 2 \sinh(3u)\cos(3v)/3,$$
$$y = 2 \sinh u \sin v + 2 \sinh(3u)\sin(3v)/3,$$
$$z = 2 \cosh(2u)\cos(2v).$$

$0 < u < 0.95,\ -\pi < v < \pi;\ -7 < x < 7,\ -7 < y < 7,\ -7 < z < 7;\ \text{viewpoint} = (4, 4, 2)$

15.2.4 Scherk's First Surface

$$z = \ln[\cos x/\cos y]$$

$-2\pi < x < 2\pi,\ -2\pi < y < 2\pi,\ -4 < z < 4;\ \text{viewpoint} = (2, -3, 2)$

15.2.5 Scherk's Second Surface

Defined parametrically by

$$x = \mathrm{Re}\{2[\ln(1 + re^{i\varphi}) - \ln(1 - re^{i\varphi})]\}$$
$$y = \mathrm{Re}[4i \arctan(re^{i\varphi})]$$
$$z = \mathrm{Re}\{2i[\ln(1 + r^2 e^{2i\varphi}) - \ln(1 - r^2 e^{2i\varphi})]\}$$

$0 < r < 1;\ 0 < \varphi < 2\pi;\ -5 < x < 5,\ -5 < y < 5,\ -5 < z < 5;\ \text{viewpoint} = (4, -7, 6)$

15.2.6 Catalan's Surface

Defined parametrically by

$$x = u - \sin u \cosh v$$
$$y = 1 - \cosh u \cosh v$$
$$z = -4 \sin(u/2) \sinh(v/2).$$

$0 < u < 4\pi,\ -2 < v < 2;\ -5 < x < 15,\ -10 < y < 10,\ -10 < z < 10;\ \text{viewpoint} = (3, -7, 2)$

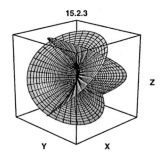

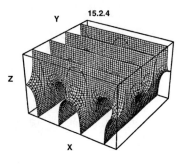

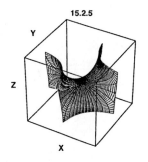

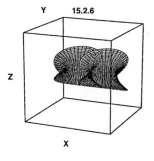

15.2.7 Richmond's Surface

Defined parametrically by

$$x = -\frac{\cos(\varphi)}{2r} - \frac{r^3\cos(3\varphi)}{6}$$

$$y = -\frac{\sin(\varphi)}{2r} - \frac{r^3\sin(3\varphi)}{6}$$

$$z = r\cos(\varphi).$$

$0.3 < r < 1.0$, $0 < \varphi < 2\pi$; viewpoint $= (7, 7, 3)$

15.2.8 Bour's Surface

Defined parametrically by:

$$x = r^2\cos(2\varphi) - \frac{r^4\cos(4\varphi)}{4}$$

$$y = -r^2\sin(2\varphi) - \frac{r^4\sin(4\varphi)}{4}$$

$$z = \frac{2r^3\cos(3\varphi)}{3}.$$

$0 < r < 1$, $0 < \varphi < 2\pi$; viewpoint $= (5\cos(\pi/3), 5\sin(\pi/3), 5)$

References

1. D.A. Hoffman, ed. *Global Theory of Minimal Surfaces: Proceedings of the Clay Mathematics Institute 2001 Summer School, Mathematical Sciences Research Institute, Berkeley, California*, in Clay Mathematics Proceedings, Vol. 2, Providence, RI: American Mathematical Society.
2. A. Gray. 1999. *Modern Differential Geometry of Curves and Surfaces with Mathematica*, 2nd Ed., Boca Raton, FL: CRC Press.

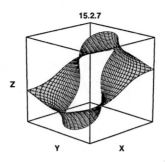

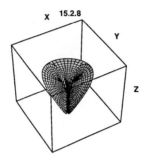

16

Regular and Semi-Regular Solids with Edges

This chapter illustrates *polyhedra* that are formed by joined, regular polygons. If all the polygons are regular and identical, the polyhedron is termed a *regular polyhedron* or regular solid. The known polyhedra are numerous (see, for instance, the comprehensive survey of Williams[1]); only the simpler and more common ones are presented here.

The projection used here is the perspective one described at the beginning of Chapter 11. The viewing point is not given for the figures here because it is not meaningful in this context; however, it is at a large distance from the figure in all cases, such that the projection is effectively a parallel one.

16.1 Platonic Solids

The five Platonic solids are also called the *regular polyhedra*. They are defined as being convex polyhedra having all faces identical, with each face being a convex regular polygon.

16.1.1   Tetrahedron ($n = 4$)

16.1.2   Hexahedron ($n = 6$, also called cube)

16.1.3   Octahedron ($n = 8$)

16.1.4   Dodecahedron ($n = 12$)

16.1.1

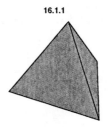

16.1.2

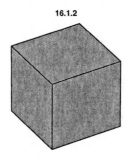

16.1.3

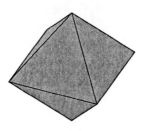

16.1.4

16.1.5   Icosahedron ($n = 20$)

───

16.2 Archimedean Solids

There are 13 Archimedean solids. They are defined as being convex polyhedra whose faces consist of two or more types of convex regular polygons.

16.2.1   Cuboctahedron

16.2.2   Great rhombicosidodecahedron

16.2.3   Great rhombicuboctahedron

16.1.5

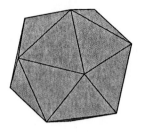

16.2.1

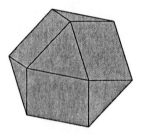

16.2.2

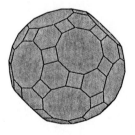

16.2.3

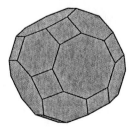

16.2.4   Icosidodecahedron

16.2.5   Small rhombicosidodecahedron

16.2.6   Small rhombicuboctahedron

16.2.7   Snub cube

16.2.4

16.2.5

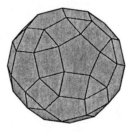

16.2.6

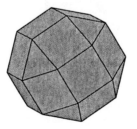

16.2.7

16.2.8   Snub dodecahedron

16.2.9   Truncated cube

16.2.10   Truncated dodecahedron

16.2.11   Truncated icosahedron

16.2.8

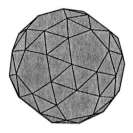

16.2.9

16.2.10

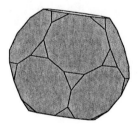

16.2.11

16.2.12   Truncated octahedron

16.2.13   Truncated tetrahedron

16.3 Duals of Platonic Solids

A *dual of a polyhedron* is made by placing a vertex at the center of the sides of the poly-hedron and placing polygons at the vertices of the polyhedron. The resulting polyhedron is the dual of the first. For Platonic solids, the dual is simply another Platonic solid. The combination of the original Platonic solid and its dual gives just three unique cases as shown here.

16.3.1   Dual of Tetrahedron = Tetrahedron

16.3.2   Dual of Hexahedron = Octahedron

16.2.12

16.2.13

16.3.1

16.3.2

16.3.3   Dual of Dodecahedron = Icosahedron

16.4 Stellated (Star) Polyhedra

A *stellated polyhedron* is formed by placing a vertex over the center of each polygon side and connecting this vertex to all the corners of the polygon. Although any convex polyhedra can be stellated, this section presents only the stellated Platonic solids. Let a be the ratio of the distance to the star vertices over the distance to the regular vertices, both measured from the center of the figure.

16.4.1   Stellated tetrahedron ($a = 1.5$)

16.4.2   Stellated hexahedron ($a = 1.5$)

16.4.3   Stellated octahedron ($a = 1.5$)

16.3.3

16.4.1

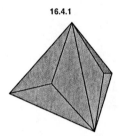

16.4.2

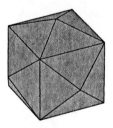

16.4.3

16.4.4   Stellated dodecahedron ($a = 1.5$)

16.4.5   Stellated icosahedron ($a = 1.3$)

Reference

1. R. Williams. 1979. *The Geometrical Foundation of Natural Structure*, New York: Dover.

16.4.4

16.4.5

17

Irregular and Miscellaneous Solids

In contrast to the previous chapter, this chapter presents solids that are not regular (or semi-regular). Smooth, closed surfaces may form part or all of the boundary of such solids such that the outward normal of the surface at any point is everywhere continuous in all directions about that point (for example, a sphere or torus). This chapter also contains a collection of surfaces with edges having no particular regularity or symmetry.

The projection used here is the perspective one described at the beginning of Chapter 11. The viewing point is not given for the figures here because it is not meaningful in this context; however, it is at a large distance from the figure in all cases, such that the projection is effectively a parallel one.

17.1   Irregular Polyhedra

17.1.1   Prism

A *prism* has identical top and bottom regular polygons and sides that meet those polygons at right angles. The regular polygon has m sides. The length of the prism's axis is given by the parameter b. Parameter a is the distance from the center to the polygonal vertices.

  a. $m=3$, $a=1.0$, $b=3.0$ (*triangular*)

  b. $m=6$, $a=1.0$, $b=3.0$ (*hexagonal*)

17.1.2   Star Prism

A *star prism* is composed of regular star polygons on the ends, with the number of star points given by m. There are then $2m$ sides joining the top and bottom. The distance from the center of the interior polygon to any vertex is given by a. The distance from the axis to any star vertex is given by the parameter c. The length of the prism's axis is given by the parameter b.

  a. $m=3$, $a=1.0$, $b=4.0$, $c=3.0$ (*triangular*)

  b. $m=6$, $a=1.0$, $b=4.0$, $c=3.0$ (*hexagonal*)

17.1.3   Anti-Prism

An *anti-prism* has identical top and bottom regular polygons of m sides, but one is rotated by $360°/(2m)$ with respect to the other. Consequently, the connection between top and bottom is composed of $2m$ identical triangles, m of which point down and m of which

17.1.1a

17.1.1b

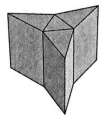

17.1.2a

17.1.2b

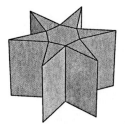

point up. The distance from the center to any vertex of the top or bottom polygon is given by a. The length of the prism's axis is given by the parameter b.

a. $m=3$, $a=1.0$, $b=3.0$ (*triangular*)

b. $m=6$, $a=1.0$, $b=3.0$ (*hexagonal*)

17.1.4 Prismoid

A *prismoid* has parallel top and bottom regular polygons, but of different size. The centers of the top and bottom both lie on an axis which is normal to them. The distance from the center to any vertex is given by a_1 and a_2 for the top and bottom, respectively. The length of the prism's axis is given by the parameter b.

a. $m=3$, $a_1=0.5$, $a_2=1.0$, $b=2.0$ (*triangular*)

b. $m=6$, $a_1=0.5$, $a_2=1.0$, $b=2.0$ (*hexagonal*)

17.1.5 Prismatoid

A *prismatoid* has parallel top and bottom regular polygons of unequal number of sides. The ratio of number of sides of the top and bottom polygons must be a whole number. The top and bottom centers lie on a single axis that is normal to both. The top and bottom polygons are assigned m_1 and m_2 sides, respectively. The distance from the center to the vertices is given by a_1 and a_2 for the top and bottom, respectively; and the length of the prism's axis is given by the parameter b. A special case, called a *cupola*, is where the sides are all either squares or equilateral triangles and all edges are of the same length a. In this case, b is

17.1.3a

17.1.3b

17.1.4a

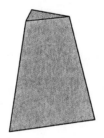

17.1.4b

constrained to the value:

$$b = a \left[1 - \frac{1}{4\sin^2(2\pi/m_1)} \right]^{1/2}.$$

a. $m_1 = 6$, $m_2 = 3$, $a_1 = 1.5$, $a_2 = 1.0$, $b = 2.0$ (*triangular–hexagonal*)

b. $m_1 = 8$, $m_2 = 4$, $a_1 = 1.5$, $a_2 = 1.0$, $b = 2.0$ (*square–octagonal*)

c. $m_1 = 3$, $m_2 = 6$, $a = 1.0$ (*cupola: triangular–hexagonal*)

d. $m_1 = 4$, $m_2 = 8$, $a = 1.0$ (*cupola: square–octagonal*)

17.1.6 Parallelepiped

A *parallelepiped* has six sides, of which opposing sides are identical and parallel. If the sides all join at right angles, it is a *right parallelepiped*; if two opposing sides meet the ends at an angle other than a right angle, it is an *oblique parallelepiped*. Let the dimensions of the ends be a by b and the remaining side length be c. Let the angle between the two faces joined

17.1.5a

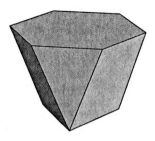

17.1.5b

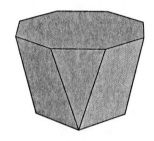

17.1.5c

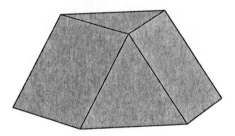

17.1.5d

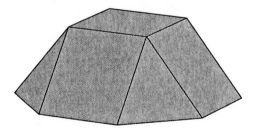

along the edges of length b be $90° \pm p°$. The remaining two faces join the ends along the edges of length a at right angles.

  a. $a=1.5$, $b=1.0$, $c=2.0$, $p=30°$ (*oblique*)

  b. $a=1.5$, $b=1.0$, $c=2.0$, $p=0°$ (*right*)

17.1.7 Pyramid

A *pyramid* has a base of m sides, with m triangles joining the base and meeting at a single apex. The pyramid is "regular" if the base is a regular polygon; in this case, the sides are all identical and the apex lies on a line that is normal to the center of the base. Irregular pyramids have irregular polygons as their bases. For regular pyramids, the length of the line joining any of the vertices of the base to its central axis is a, and the height of the pyramid is b.

  a. $m=3$, $a=1.0$, $b=2.0$ (*triangular*)

  b. $m=6$, $a=1.0$, $b=2.0$ (*hexagonal*)

17.1.6a

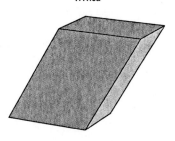

17.1.6b

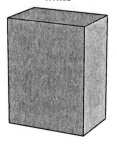

17.1.7a

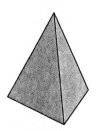

17.1.7b

17.1.8   Dipyramid

A *dipyramid* is simply two pyramids joined base-to-base. The length of the line joining any of the vertices of the mutual base to its center is a, and the height of each pyramid is b. The number of sides of each pyramid is given by m. As for pyramids, the figure is regular if the base is a regular polygon and irregular if the base is not.

   a. $m=3$, $a=1.0$, $b=1.5$ (*triangular*)

   b. $m=6$, $a=1.0$, $b=1.5$ (*hexagonal*)

17.1.9   Trapezohedron

A *trapezohedron* has n sides extending from each of two apices. It is similar to the dipyramid, but here each side is a rhombus rather than a triangle. Like the dipyramid, it can be constructed from two identical parts, except for suitable translations and rotations. The length of the line perpendicular to the central axis and connecting to the vertices is a, and the length of the line from the outermost points to the nearest apex is b. The number of sides of either of the top or bottom part is m.

   a. $m=3$, $a=1.0$, $b=1.5$ (*triangular*)

   b. $m=6$, $a=1.0$, $b=1.5$ (*hexagonal*)

17.1.8a

17.1.8b

17.1.9a

17.1.9b

17.1.10   Obelisk

An *obelisk* has parallel, but non-congruent, top and bottom rectangles. The centers of the top and bottom both lie on an axis that is normal to them and are separated by the distance h. The top rectangle has dimensions of a by b while the bottom one has dimensions of c by d. Trapezoids then connect the top and bottom.

a. $a=0.5$, $b=2.0$, $c=2.0$, $d=3.0$, $h=1.0$

b. $a=0.5$, $b=1.0$, $c=3.0$, $d=2.0$, $h=3.0$

17.1.11   Irregular Dodecahedron

The *irregular dodecahedron* has parallel squares at the top and bottom. The sides are composed of ten additional quadrilaterals. Eight of these are identical, except for rotation and translation; and the two remaining ones are identical. Let the distance from the center of the square to its vertices be a, the axial distance between top and bottom be c, and the lateral extension of the middle of the figure beyond a be b. Note that for $b=0$ the figure degenerates to a right parallelepiped.

a. $a=0.5$, $b=0.2$, $c=1.5$

b. $a=0.5$, $b=1.0$, $c=1.5$

17.1.10a

17.1.10b

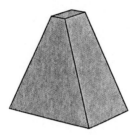

17.1.11a

17.1.11b

17.2 Miscellaneous Closed Surfaces with Edges

17.2.1 Cylinder

The axial length of the *cylinder* is given by b, and the radius is given by a. An oblique cylinder is displayed for $c > 0$, where c is the relative offset of the top and bottom. When $b \ll a$, the figure is commonly called a *disk*.

a. $a = 1.0$, $b = 3.0$, $c = 0.0$ (*right circular*)

b. $a = 1.0$, $b = 3.0$, $c = 0.5$ (*oblique circular*)

c. $a = 1.0$, $b = 0.2$, $c = 0.0$ (*right circular*)

17.2.2 Cone

The height of the *cone* is b, and the radius of the base is given by a. An oblique cone is formed when $c > 0$, where c is the lateral displacement of the apex from the axis normal to the center of the base.

a. $a = 1.0$, $b = 3.0$, $c = 0.0$ (*right circular*)

17.2.1a

17.2.1b

17.2.1c

17.2.2a

b. $a=1.0$, $b=3.0$, $c=0.7$ (*oblique circular*)

17.2.3 Frustrum of a Cone

A *frustrum* is a cone truncated by a plane. The height of the frustrum is b, the radius of the base is given by a, and the radius of the top is given by c.

$a=2.0$, $b=2.0$, $c=1.0$

17.2.4 Hemisphere

A *hemisphere* is one-half of a sphere. It can be generalized to a hemi-ellipsoid with separate parameters (a,b,c) for the lengths of the three axes (x,y,z) of the ellipsoid. Using $a=b=c$ gives the exact hemisphere.

$a=1.0$, $b=1.0$, $c=1.0$

17.2.5 Rectangular Torus

A *rectangular torus* is a torus whose cross-section is rectangular rather than circular. The outside radius is given by a and the inside radius is given by b ($b<a$). The height of a vertical cross-section is given by c.

$a=1.0$, $b=0.5$, $c=0.5$

17.2.2b

17.2.3

17.2.4

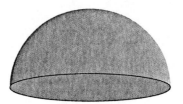

17.2.5

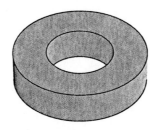

Index